STUDENT STUDY GUIDE & SELECTED SOLUTIONS MANUAL

DAVID D. REID

UNIVERSITY OF CHICAGO

PHYSICS

THIRD EDITION VOLUME II

JAMES S. WALKER

Upper Saddle River, NJ 07458

Assistant Editor: Jessica Berta
Project Manager: Christian Botting
Senior Editor: Erik Fahlgren
Editor-in-Chief, Science: Dan Kaveney
Executive Managing Editor: Kathleen Schiaparelli
Senior Managing Editor: Nicole M. Jackson
Assistant Managing Editor: Karen Bosch Petrov
Production Editor: Jessica Barna
Supplement Cover Manager: Paul Gourhan
Supplement Cover Designer: Christopher Kossa
Manufacturing Buyer: Ilene Kahn
Manufacturing Manager: Alexis Heydt-Long
Cover Images: Three rows of organ pipes: Kenneth Garrett/National Geographic/ Getty Images, Inc.; opera singer in a Viking helmet: C Squared Studios/Photodisc Green/Getty Images, Inc.; magnet covered in iron fillings: Steve Taylor/Stone/Getty Images, Inc.; MRI scan of human head and shoulders (Digital Composite): David Job/Stone/Getty Images, Inc.

Pearson Prentice Hall
Pearson Education, Inc.
Upper Saddle River, NJ 07458

Printed in the United States of America

10 9 8 7 6 5 4 3 2 1

ISBN 0-13-236963-X

Pearson Education Ltd., *London*
Pearson Education Australia Pty. Ltd., *Sydney*
Pearson Education Singapore, Pte. Ltd.
Pearson Education North Asia Ltd., *Hong Kong*
Pearson Education Canada, Inc., *Toronto*
Pearson Educación de Mexico, S.A. de C.V.
Pearson Education—Japan, *Tokyo*
Pearson Education Malaysia, Pte. Ltd.

To my wife, *Annie*

PREFACE

This study guide is designed to assist you in your study of the fascinating and sometimes challenging world of physics using *Physics*, *Third Edition* by James S. Walker. To do this I have provided a Chapter Review, which consists of a comprehensive (but brief) review of almost every section in the text. Numerous solved examples and exercises appear throughout each Chapter Review. The examples follow the two-column format of the text, while the solutions to the exercises have a more traditional layout. Together with the Chapter Review, each chapter contains a list of objectives, a practice quiz, a glossary of key terms and phrases, a table of important formulas, and a table that reviews the units and dimensions of the new quantities introduced.

In addition to the above materials that I have provided, you will also find Warm-Up and Puzzle questions by Just in Time Teaching innovators Gregory Novak and Andrew Gavrin (Indiana University-Purdue University, Indianapolis), Answers to Selected Conceptual Questions and Exercises, and Solutions to Selected End-of-Chapter Problems from the *Instructor's Solutions Manual*. Taken together, the information in this study guide, when used in conjunction with the main text, should enhance your ability to master the many concepts and skills needed to understand physics, and therefore, the world around you. Work hard, and most importantly, have fun doing it!

I am indebted to many for helping me to complete this work. Most directly, I thank the Prentice Hall staff for their work with me on this volume. I also wish the acknowledge Dr. Anand P. Batra of Howard University. He provided an excellent review of the physics content for the first edition of this Study Guide and made countless valuable suggestions.

David D. Reid
University of Chicago
July 2006

TABLE OF CONTENTS

CHAPTER 19
ELECTRIC CHARGES, FORCES, and FIELDS

Chapter Objectives

After studying this chapter, you should

1. know the different types of electric charge and the magnitude of the smallest available charge.
2. know the difference between insulators and conductors.
3. be able to use Coulomb's law and apply it to situations involving many charges.
4. know the definition of, and basic uses for, the electric field.
5. be able to sketch electric field lines.
6. know how electrostatic shielding works.
7. be able to use Gauss's law to find the electric field in certain situations.

Warm-Ups

1. Can there be an electric field at a point where there is no charge? Can there be a charge at a place where there is no field? Please write a one- or two-sentence answer to each of these questions.
2. Let's say you are holding two tennis balls (one in each hand), and let's say these balls each have a charge Q. Estimate the maximum value of Q so that the balls do not repel each other so hard that you can't hold on to them.
3. Is it possible for two electric field lines to cross? If so, under what conditions? Do electric field lines ever end? If so, under what conditions?
4. Near its surface, the earth has an electric field that points straight down and has a magnitude of about 150 N/C. Estimate the charge that you would have to place on a basketball so that the electric force on the ball would balance its weight.

Chapter Review

This chapter begins our study of electricity and magnetism. Electric and magnetic forces are due to a phenomenon that we have not studied in previous chapters, namely, electric charge. The existence of electric charge and its consequences creates a completely new branch of physics that adds to the concepts and phenomena that we studied in previous chapters.

19–1 – 19–2 Electric Charge, Insulators, and Conductors

Through a series of experiments over many years we have now come to understand that we are all made up of smaller objects (atoms) that contain **electric charge**. To the best of our present knowledge, electric charge is a fundamental property of nature that comes in two types, called **positive charge** and **negative charge**. Most bulk matter contains an equal number of positive and negative charges and is said to be electrically **neutral** (or to have zero net charge).

Atoms consist of a small central nucleus that contains positively charged particles called **protons**. The nucleus is surrounded by an equal number of negatively charged particles called **electrons**. The electric charge of protons and electrons has the same magnitude

$$e = 1.60 \times 10^{-19}\ \mathrm{C}$$

Electrons have a charge of $-e$, and protons carry a charge of $+e$. The SI unit of electric charge is called the **coulomb** (C). This is a new unit that is not derived from just [L], [M], and [T]. However, the coulomb is officially considered to be a derived unit; we will see its derivation in a later chapter. For now we will denote the dimension associated with electric charge as [C].

One of the properties of electric charge is that it is **quantized**. This fact means that charge comes only in discrete units. The smallest available charge is that of a proton or electron, e. Another property of electric charge is that it is conserved. Therefore, *in any physical process, electric charge is never created or destroyed; the total electric charge of the universe remains constant.*

The positive and negative charges in an object can become separated, usually by the movement of electrons, so that one side of the object contains more of the negative charge, and the other side is left with more of the positive charge. Such objects are said to be **polarized**. In atoms and molecules (bound groups of atoms) electrons can be completely removed, or extra electrons can be added. An atom with one or more electrons removed will have a net positive charge and is called a **positive ion**; an atom with extra electrons will have a net negative charge and is called a **negative ion**.

Important to our ability to make practical use of electricity is the fact that in some materials, called **conductors**, electrons are relatively free to move, while in other materials, called **insulators**, electrons are not very free to move. Metals are typically good conductors of electricity. Examples of commonly used insulators are rubber, plastic, and wood. There are also materials, called **semiconductors**, whose behavior is not clearly conducting or insulating. These materials can be manipulated to be more conducting in some situations and more insulating in others. The ability to manipulate semiconductors to control the flow of electricity has been a major triumph of modern technology.

Practice Quiz

1. Which of the following values is *not* a possible charge on an ion?

(a) 1.60×10^{-19} C **(b)** 3.20×10^{-19} C **(c)** -1.60×10^{-19} C

(d) 4.80×10^{-19} C **(e)** -2.80×10^{-19} C

19–3 Coulomb's Law

Electric charges exert forces on one another. The rule for determining this force is called **Coulomb's law**. This law states that for two stationary point charges of magnitudes q_1 and q_2, the magnitude of the electrostatic force between them is proportional to the product $q_1 \cdot q_2$ and inversely proportional to the square of the distance r between them. Therefore, the magnitude of the force is given by

$$F = k\frac{q_1 q_2}{r^2}$$

where k is a constant of proportionality that has the value

$$k = 8.99 \times 10^9 \text{ N} \cdot \text{m}^2/\text{C}^2$$

Two additional rules allow us to determine the direction of the force. One rule is that *the force is directed along the line joining the charges*. The second rule is that *like charges repel each other and opposite charges attract each other*. Note that there are actually two forces, one exerted on q_1 by q_2 ($\vec{\mathbf{F}}_{12}$) and one exerted on q_2 by q_1 ($\vec{\mathbf{F}}_{21}$). These forces have equal magnitudes, given by F, and point in opposite directions in accordance with Newton's law of action and reaction. When more than two charges are involved, the force on any one charge can be determined using the principle of **superposition**. This principle states that the force on any charge is the vector sum of the forces that each of the other charges exerts on it individually.

In addition to individual point charges, it is also common to deal with continuous distributions of charge. An interesting example is that of an amount of charge Q distributed uniformly over the surface of a sphere. The force experienced by a point charge q outside the sphere works out, by superposition, to look just like Coulomb's law:

$$F = k\frac{qQ}{r^2}$$

where r, in this case, is the distance of q from the center of the sphere. When dealing with continuous distributions of charge, it is often convenient to work with charge densities. In the case at hand, we have a

surface charge density, σ, which is the charge per area ($\sigma = Q/A$) over the sphere. Thus, for an amount of charge Q spread over the surface of a sphere of radius R, the surface charge density is given by

$$\sigma = \frac{Q}{4\pi R^2}$$

and its SI unit is C/m^2.

Example 19–1 Coulomb's Law and Superposition Given a configuration of three charged particles such that charge $q_1 = 5.3$ nC has (x, y) coordinates of (1.5 m, 0.0 m), charge $q_2 = 9.2$ nC is located at (0.0 m, 2.0 m), and charge $q_3 = -2.4$ nC is located at the origin, determine the net force on charge q_3.

Picture the Problem The diagram shows the charge configuration of q_1, q_2, and q_3.

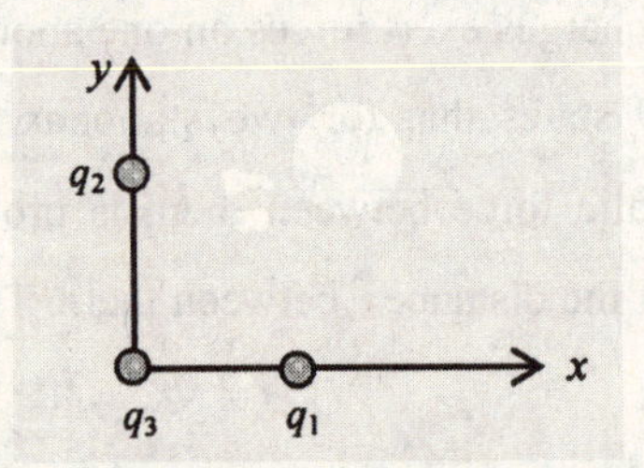

Strategy Our basic strategy, in keeping with the principle of superposition, is to determine the force on q_3 due to q_1 and q_2 separately using Coulomb's law, and then calculate the net force on q_3 as the vector sum of the two.

Solution

1. From the given coordinates, the distance between q_1 and q_3 is: $r_{13} = 1.5$ m

2. Using Coulomb's law, the magnitude of the force on q_3 due to q_1 is:

$$F_{31} = \frac{k|q_1 q_3|}{r_{13}^2}$$

$$= \frac{(8.99\times10^9\ \text{N}\cdot\text{m}^2/\text{C}^2)|(5.3\times10^{-9}\ \text{C})(-2.4\times10^{-9}\ \text{C})|}{(1.5\ \text{m})^2}$$

$$= 5.1\times10^{-8}\ \text{N}$$

3. Since the charges have opposite signs, $\vec{\mathbf{F}}_{31}$ must point toward q_1: $\vec{\mathbf{F}}_{31} = (5.1\times10^{-8}\ \text{N})\hat{\mathbf{x}}$

4. From the given coordinates, the distance between q_2 and q_3 is: $r_{23} = 2.0$ m

5. Using Coulomb's law, the magnitude of the force on q_3 due to q_2 is:

$$F_{32} = \frac{k|q_2 q_3|}{r_{23}^2}$$
$$= \frac{(8.99\times10^9\ \text{N}\cdot\text{m}^2/\text{C}^2)|(9.2\times10^{-9}\ \text{C})(-2.4\times10^{-9}\ \text{C})|}{(2.0\ \text{m})^2}$$
$$= 5.0\times10^{-8}\ \text{N}$$

6. Since the charges have opposite signs, $\vec{\mathbf{F}}_{31}$ must point toward q_2:

$$\vec{\mathbf{F}}_{32} = (5.0\times10^{-8}\ \text{N})\hat{\mathbf{y}}$$

7. The net force on q_3 is the sum of the two forces:

$$\vec{\mathbf{F}}_3 = \vec{\mathbf{F}}_{31} + \vec{\mathbf{F}}_{32} = (5.1\times10^{-8}\ \text{N})\hat{\mathbf{x}} + (5.0\times10^{-8}\ \text{N})\hat{\mathbf{y}}$$

Insight Notice that because q_3 is a negative charge, we used the absolute value of the product of the charges in each application of Coulomb's law.

Practice Quiz

2. If two oppositely charged particles are placed side by side, which charge experiences the greater magnitude of force?

   (a) the positive charge

   (b) the negative charge

   (c) neither charge experiences any force

   (d) they experience forces of equal magnitude

   (e) none of the above

3. If two oppositely charged particles are placed side by side with the positive charge to the left of the negative charge, what is the direction of the force on the positive charge?

   (a) to the left

   (b) to the right

   (c) perpendicular to the line joining them

   (d) there is no force on the positive charge

   (e) none of the above

4. A point charge of $+Q$ placed a distance d away from the origin on the positive x-axis applies a force $\vec{\mathbf{F}}$ on a charge q at the origin. If another charge of $-Q$ is placed a distance d away from the origin on the negative x-axis, what will be the net force on q?

   (a) $-\vec{\mathbf{F}}$ **(b)** $\vec{\mathbf{F}}$ **(c)** $-2\vec{\mathbf{F}}$ **(d)** $2\vec{\mathbf{F}}$ **(e)** 0

19–4 – 19–5 The Electric Field and Electric Field Lines

In considering Coulomb's law, you may have noticed that the two charges apply forces to each other even though they are not in physical contact. It is useful, and ultimately very important, to define an intermediary for the electrostatic force. This intermediary is called the **electric field**. We say that every charge creates an electric field in the space around it and that this electric field exerts a force on other charges placed within it. The electric field $\vec{\mathbf{E}}$ is a vector quantity defined as the force per unit of positive charge at a given location:

$$\vec{\mathbf{E}} = \vec{\mathbf{F}}/q_0$$

In the above equation, q_0 (imagined to be placed at the given location) represents a *test charge,* which is a positive charge so small in magnitude that its presence does not significantly alter the electric field in the region. The electric field is independent of q_0. The SI unit of electric field is N/C. Notice from the above definition that the electric field is defined in terms of the force on a positive charge. Therefore, a positive charge experiences a force in the direction of $\vec{\mathbf{E}}$, and a negative charge experiences a force in the opposite direction of $\vec{\mathbf{E}}$.

The electric field due to a point charge can easily be determined using Coulomb's law with one charge q as the source of the field and the other charge as a test charge. This configuration reveals that the magnitude of the electric field a distance r away from a point charge is given by

$$E = k\frac{q}{r^2}$$

To determine the electric field, at a given location, due to multiple charges, we use the principle of superposition. Each point charge contributes to the total electric field vector according to the above equation for the magnitude and the rule that electric fields point in the direction of the force on a positive charge. The total electric field is found by taking the vector sum of the electric fields due to each charge.

Example 19–2 Electric Fields and Superposition A configuration of three charges whose values are $q_1 = q_2 = -q_3 = 3.88\ \mu\text{C}$ is shown below. Charge q_1 is located at (–3.00 m, –3.00 m), charge q_2 is located at (–3.00 m, 3.00 m), and charge q_3 is located at (3.00 m, 3.00 m). **(a)** Determine the magnitude and direction of the electric field at the origin. **(b)** Determine the magnitude and direction of the force on a –1.00-nC charge placed at the origin.

Picture the Problem The sketch shows the charge configuration.

y
q_2 q_3
x
q_1

Strategy For part (a), we use superposition to find the electric field by adding the electric fields due to each charge. For part (b), we use the definition of electric field to determine the force.

Solution

Part (a)

1. From the given coordinates, the squared distances of the charges from the origin are: $r_1^2 = (-3.00\text{ m})^2 + (-3.00\text{ m})^2 = 18\text{ m}^2 = r_2^2 = r_3^2$

2. Because all the charges have equal magnitudes and are at equal distances from the origin, the magnitude of the field due to each charge is: $E_1 = E_2 = E_3 = \dfrac{kq}{r^2} = \dfrac{(8.99\times10^9\text{ N}\cdot\text{m}^2/\text{C}^2)(3.88\times10^{-6}\text{C})}{18\text{ m}^2} = 1938\text{ N/C}$

3. Each charge is an equal distance from both the x- and y-axes. Because electric fields point radially away from positive charges and radially toward negative charges, each field makes a 45° angle with the x- and y-axes:

$$\sin(45^\circ) = \cos(45^\circ) = 1/\sqrt{2} \;\therefore$$
$$\vec{\mathbf{E}}_1 = \frac{1938\text{ N/C}}{\sqrt{2}}[\hat{\mathbf{x}} + \hat{\mathbf{y}}]$$
$$\vec{\mathbf{E}}_2 = \frac{1938\text{ N/C}}{\sqrt{2}}[\hat{\mathbf{x}} - \hat{\mathbf{y}}]$$
$$\vec{\mathbf{E}}_3 = \frac{1938\text{ N/C}}{\sqrt{2}}[\hat{\mathbf{x}} + \hat{\mathbf{y}}]$$

4. The net electric field then is:

$$\vec{\mathbf{E}} = \vec{\mathbf{E}}_1 + \vec{\mathbf{E}}_2 + \vec{\mathbf{E}}_3 = \frac{1938\text{ N/C}}{\sqrt{2}}(3\hat{\mathbf{x}} + \hat{\mathbf{y}}) = (4111\text{ N/C})\hat{\mathbf{x}} + (1370\text{ N/C})\hat{\mathbf{y}}$$

5. The magnitude of the electric field is:

$$E = \left(E_x^2 + E_y^2\right)^{1/2} = \left[(4111\text{ N/C})^2 + (1370\text{ N/C})^2\right]^{1/2} = 4330\text{ N/C}$$

6. The direction of the electric field is:

$$\theta_E = \tan^{-1}\left(\frac{E_y}{E_x}\right) = \tan^{-1}\left(\frac{1370\text{ N/C}}{4111\text{ N/C}}\right) = 18.4^\circ$$

Part (b)

7. By definition, the magnitude of the force on a charge in an electric field is: $F = |q|E = |-1.00\times10^{-9}\text{ C}|(4333\text{ N/C}) = 4.33\times10^{-6}\text{ N}$

8. Because the charge in question is negative, the

direction of the force on it must be opposite that of the electric field: $\theta_F = 18.4^\circ + 180^\circ = 198^\circ$

Insight Be absolutely sure you understand how the result in step 3 relates to the position of the charges and the fact that the fields of point charges are radial.

A visual representation of electric fields can be obtained by drawing **electric field lines**, see Figure 19–15 on page 644 in the text. The electric field lines can be drawn by following four rules:

1. Electric field lines are tangent to the electric field at every point.
2. Electric field lines start on positive charge (or at infinity).
3. Electric field lines end on negative charge (or at infinity).
4. The number of electric field lines is proportional to the magnitude of the source charge.

Exercise 19–3 Electric Field Lines **(a)** Sketch the electric field lines of a point charge of positive charge Q. **(b)** Sketch the electric field lines for a point charge of charge $-2Q$.

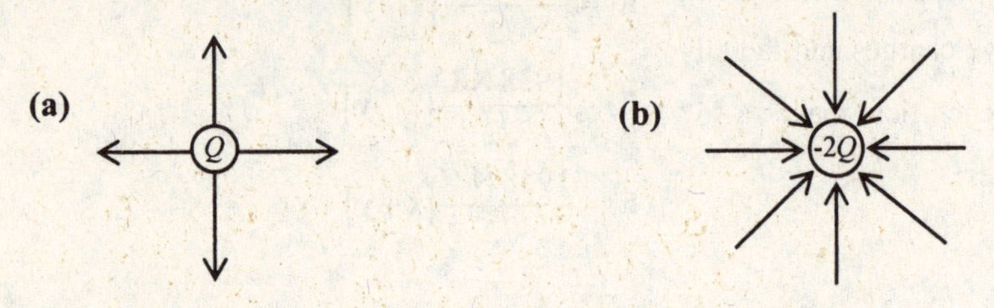

There are only two significant differences between the two sketches. (1) The field lines point away from the positive charge and toward the negative charge. (2) Twice as many lines are drawn for the charge with twice the magnitude.

Practice Quiz

5. The magnitude of the electric field a distance of 5.28×10^{-11} m away from a proton is

   (a) 8.28×10^{-8} N/C **(b)** 27.3 N/C **(c)** 57.5 N/C

   (d) 5.16×10^{11} N/C **(e)** none of the above

6. If, in a scaled drawing, 4 electric field lines are drawn around a point charge q_1 and 12 electric field lines are drawn around a different point charge q_2, this indicates that

   (a) the magnitude of q_1 is 12 times that of q_2.

(b) the magnitude of q_2 is 1/4 that of q_1.

(c) the magnitude of q_2 is 3 times that of q_1.

(d) q_1 and q_2 have equal magnitudes.

(e) No statement can be made from this information.

19–6 Shielding and Charging by Induction

This section collects several useful facts related to the electrostatics of conductors. The first of these facts states what happens to excess charge on a conductor:

> *In electrostatic equilibrium, any excess charge on a conductor, whether positive or negative, must lie on the surface of the conductor.*

The next fact dealing with conductors has to do with the phenomenon called **electrostatic shielding**:

> *In electrostatic equilibrium, the electric field within the body of a conducting material must be zero.*

This fact remains true even if the conductor sits in an externally applied electric field. In this sense, the inside of the conductor is shielded from the electric field.

The fact that charges are relatively free to move in conductors makes it possible for conductors to have a net positive or negative charge through a process called **charging by induction**. In this process, a charged object (with charge Q) is brought near a neutral conductor. Free electrons in the conductor will be attracted to the side of the conductor nearer to, or pushed to the side further from, the charged object (depending on the sign of Q). If the conductor is then **grounded**, (that is, connected to an object like Earth that can give or receive a large supply of electrons) free electrons can either flow off or onto the conductor, leaving it with a net electric charge once the conductor is disconnected from the ground. The sign of the net charge on the conductor will be opposite that of the object used to induce the charge.

Practice Quiz

7. If a conductor is placed in an external electric field, the electric field inside the body of the conductor

   (a) always equals zero.

   (b) never equals zero.

   (c) is nonzero only when electrostatic equilibrium is reached.

   (d) is zero only if it carries excess charge.

   (e) None of the above.

19–7 Electric Flux and Gauss's Law

In this section, we learn what some consider to be the fundamental law of electrostatics, **Gauss's law**. (We'll see later that this law is also important for electrodynamics as well.) Important to understanding this law is the concept of **electric flux**, which we review first.

In any region of space, you can imagine a surface (or a membrane) of area A. An electric field $\vec{\mathbf{E}}$, assumed to be uniform in that region, may flow through this surface. The flow of the electric field through a surface is represented by the electric flux, Φ, of $\vec{\mathbf{E}}$ through this surface. The extent to which $\vec{\mathbf{E}}$ flows through the surface depends on the angle, θ, between the electric field vector and the direction normal to the surface. Specifically, we have that

$$\Phi = EA\cos\theta$$

Notice that the flux is not a vector quantity and that its SI unit is $\text{N}\cdot\text{m}^2/\text{C}$. If the surface is closed, such as the surface of a sphere, then the following sign convention for the flux must be used:

- Φ is positive for electric field lines that leave the enclosed volume.
- Φ is negative for electric field lines that enter the enclosed volume.

Example 19–4 Electric Flux A uniform electric field of magnitude 250 N/C makes an angle of 35° with the normal to a circular aperture of radius 65 cm. If the aperture lies entirely in the electric field, what is the electric flux through the aperture?

Picture the Problem The perspective sketch shows a circular aperture, the electric field vector, and a vector, $\hat{\mathbf{n}}$, representing the direction normal to the surface.

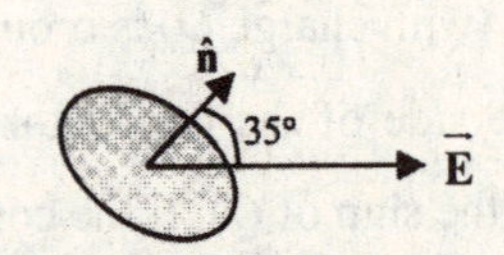

Strategy The expression for the electric flux through an area, A, can be used directly for this calculation. We need only to first calculate the area of the aperture.

Solution

1. The area of the circular aperture is: $A = \pi r^2$

2. The electric flux then is: $\Phi = E(\pi r^2)\cos\theta = (250\ \text{N/C})\pi(0.65\ \text{m})^2\cos(35°)$
$= 270\ \text{N}\cdot\text{m}^2/\text{C}$

Insight The main part of this problem is getting the picture correct. Once you correctly identify the situation being described the mathematical calculation is straightforward.

Charged objects create electric fields. Any surface that encloses a charged object will, therefore, have an electric flux through it. Gauss's law relates the electric flux through the surface to the enclosed charge in the following way:

If a charge q is enclosed by an arbitrary surface, the electric flux through the surface is $\Phi = q/\varepsilon_0$.

The constant ε_0 is called the **permittivity of free space**; its value is

$$\varepsilon_0 = 8.85 \times 10^{-12} \frac{\text{C}^2}{\text{N}\cdot\text{m}^2}$$

and is related to the constant k in Coulomb's law by $k = 1/4\pi\varepsilon_0$.

Gauss's law can be very useful for calculating electric fields in certain cases where the source of the field is highly symmetric. When we enclose a charge with an imaginary surface for the purpose of using Gauss's law, we call this surface a **Gaussian surface**. For point charges and spherically symmetric charge distributions, take the Gaussian surface to be a spherical surface centered on the charge (or center of the charge distribution) in question. For a flat sheet of charge, take the Gaussian surface to be a cylindrical surface with its axis perpendicular to the sheet (the only nonzero flux will be through the top and bottom edges). For linear or cylindrical charge distributions, also take the Gaussian surface to be a cylinder (the only nonzero flux will be through the sides).

Example 19–5 Gauss's Law An infinite thin sheet of charge has a uniform surface charge density of $-7.2\ \text{nC/m}^2$. Charge configurations of this type create uniform electric fields on each side of the sheet. Determine the electric field created by this sheet of charge.

Picture the Problem The sketch shows a section of the infinite sheet of charge, some of the electric field lines, and a cylindrical Gaussian surface that will be used to solve this problem.

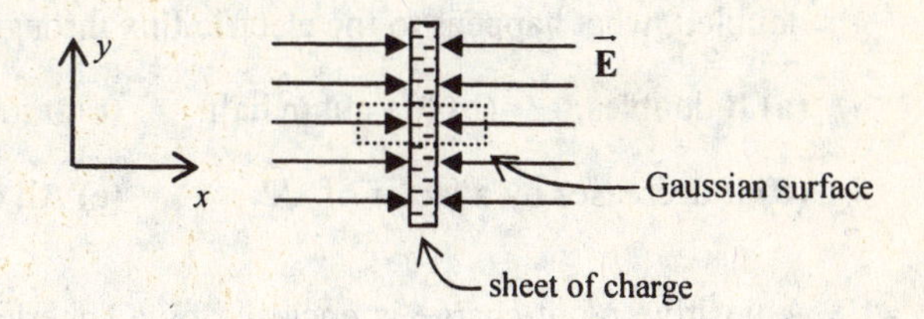

Strategy We apply Gauss's law to the cylindrical Gaussian surface of radius r to determine the electric field.

Solution

1. The only nonzero contributions to the electric flux through the Gaussian cylinder come from the two circular ends: $\Phi = -2EA = -2E\pi r^2$

2. Determine the charge enclosed by the Gaussian surface from the surface charge density: $\sigma = q/A \Rightarrow q = \sigma A = \sigma\pi r^2$

3. From Gauss's law we have: $\Phi = q/\varepsilon_0 \Rightarrow -2E\pi r^2 = \sigma\pi r^2/\varepsilon_0 \Rightarrow E = -\sigma/2\varepsilon_0$

$$\therefore \quad E = -\frac{-7.2\times10^{-9}\ \text{C/m}^2}{2\left(8.85\times10^{-12}\ \text{C}^2/\text{N}\cdot\text{m}^2\right)} = 410\ \text{N/C}$$

4. The electric field on the right side is: $\vec{\mathbf{E}}_{\text{right}} = -(410\ \text{N/C})\hat{\mathbf{x}}$

5. The electric field on the left side of the sheet is: $\vec{\mathbf{E}}_{\text{left}} = (410\ \text{N/C})\hat{\mathbf{x}}$

Insight Notice that the electric field is discontinuous at the surface. You can see this same effect in Active Example 19–3 on page 652 in the text. Can you explain why the results are different for this Example and the one in the text?

Practice Quiz

8. An electric field is directed along the normal to a surface that lies in the field. If the area of the surface is doubled, what happens to the electric flux through the surface?

   (a) It doubles. **(b)** It's cut in half. **(c)** It increases by a factor of $\sqrt{2}$.
   (d) It decreases by a factor of $\sqrt{2}$. **(e)** It stays the same.

9. An electric field is directed at 90° to the normal to a surface that lies in the field. If the electric field is doubled, what happens to the electric flux through the surface?

   (a) It doubles. **(b)** It's cut in half. **(c)** It increases by a factor of $\sqrt{2}$.
   (d) It decreases by a factor of $\sqrt{2}$. **(e)** All of the above.

10. A positive point charge is enclosed by a spherical surface of radius *r* producing an electric flux, Φ, through this surface. If this same charge is enclosed by a spherical surface with a radius of 2*r*, the electric flux through this larger surface will be

    (a) Φ. **(b)** 2Φ. **(c)** 4Φ. **(d)** Φ^2. **(e)** zero.

Reference Tools and Resources

I. Key Terms and Phrases

electric charge a property of particles that is important to the structure of atoms and molecules and acts as the source of a fundamental force of nature

protons positively charged particles, in nuclei, that possess the smallest measurable electric charge

electrons negatively charged particles found in atoms that posses the smallest measurable electric charge

conductors materials, such as metals, in which electrons easily flow

insulators materials, such as rubber, in which electrons do not readily flow

semiconductors materials with electrical properties intermediate between conductors and insulators

Coulomb's law the force law between two point charges

superposition the principle that electric forces and fields, that result from multiple sources, are obtained by vector addition of the results from the individual sources

charge density a measure of the compactness of electric charge that is especially useful with continuous charge distributions

electric field the electric force per unit of positive charge in space

electric field lines a pictorial representation of an electric field

electrostatic shielding the phenomenon that the electric field inside a conductor in static equilibrium is zero even when the conductor is exposed to an external electric field

electric flux a measure of the extent to which an electric field flows through an area

Gauss's law the law that relates the electric flux through a closed surface to the net charge it encloses

permittivity of free space the fundamental constant of electrostatics

Gaussian surface an imaginary surface used for applying Gauss's law

II. Important Equations

Name/Topic	Equation	Explanation
Coulomb's law	$F = k\frac{q_1 q_2}{r^2}$	The force between two point charges
electric field	$\vec{\mathbf{E}} = \vec{\mathbf{F}}/q_0$	The electric field is the force per unit positive charge
electric flux	$\Phi = EA\cos\theta$	Electric flux measures the extent to which the electric field flows through an area
Gauss's law	$\Phi = q/\varepsilon_0$	Gauss's law relates electric charge to the flux generated by its electric field

III. Know Your Units

Quantity	Dimension	SI Unit
electric charge (q)	[C]	C
electric field ($\vec{\mathbf{E}}$)	$[M][L][T^{-2}][C^{-1}]$	N/C
electric flux (Φ)	$[M][L^3][T^{-2}][C^{-1}]$	$\frac{N \cdot m^2}{C}$
permittivity of free space (ε_0)	$[C^2][T^2][M^{-1}][L^{-3}]$	$\frac{C^2}{N \cdot m^2}$

Puzzle

SYMMETRY

Consider the figure. The positively charged balls are located at (-1,1,0) and (1,-1,0) and each carries a charge $+Q$. The negatively charged balls are located at (1,1,0) and (-1,-1,0), and each carries a charge $-Q$. There are three specially marked points: O is the origin, $P1$ is (0,0,2), and $P2$ is (2,0,0). There are no charges at these points. Rank the points O, $P1$, and $P2$ in order of decreasing electric field strength. The correct answer is:

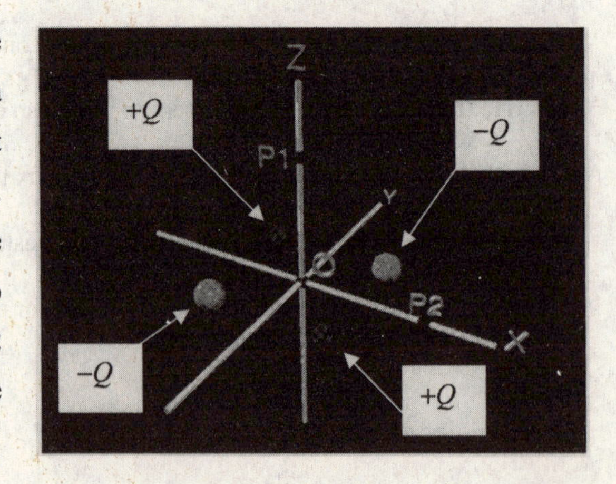

1. $P1 > P2 > O$ 2. $O > P1 > P2$ 3. $P1 > O > P2$ 4. $O > P2 > P1$ 5. $O > P2 = P1$
6. $P2 > P1 = O$ 7. $P2 > P1 > O$ 8. $P1 = P2 > O$
9. The fields are all equal.
10. None of the above

Answers to Selected Conceptual Questions and Exercises

Conceptual Questions

4. Initially, the bits of paper are uncharged and are attracted to the comb by polarization effects. (See Figure 19-5 and the accompanying discussion.) When one of the bits of paper comes into contact with the comb, it acquires charge from the comb. Now the piece of paper and the comb have charge of the same sign, and hence there is a repulsive force between them.

6. No. Even uncharged objects are attracted to a charged rod, due to polarization effects. See Figure 19-5 and the accompanying discussion.

10. The proton can move in any direction at all relative to the direction of the electric field. On the other hand, the direction of the proton's acceleration *must* be in the same direction as the electric field.

Conceptual Exercises

6. From Coulomb's law (Equation 19-5), it is clear that the magnitude of the force exerted on each of the charges is proportional to the magnitude of its charge, and inversely proportional to the square of the distance from the charge to the center of the sphere. The only exception is charge 1, which is inside the sphere and hence experiences zero net force. Combining these considerations yields the following ranking in order of increasing magnitude of force: charge 1 < charge 2 < charge 4 < charge 3 < charge 5.

10. **(a)** Sphere A is strongly attracted to sphere D, and hence it has the charge $-Q$. Sphere A would also be attracted to sphere D if its charge were zero, but in that case the deflection of the sphere would be smaller, as in the case of spheres C and D. **(b)** Sphere B is repelled by sphere D, and hence its charge must be positive. It follows that it has the charge $+Q$. **(c)** Sphere C is only weakly attracted to spheres B and D. Compare, for example, the attraction between spheres A and D and spheres C and D. It follows that sphere C has a charge of zero.

24. The electric flux through a closed surface is directly proportional to the net charge it encloses. Thus, the ranking of the Gaussian surfaces in order of increasing electric flux is as follows: D < C < B < A. To be specific, the electric flux for these surfaces is $\Phi_D = -q/\varepsilon_0$, $\Phi_C = 0$, $\Phi_B = +q/\varepsilon_0$, and $\Phi_A = +2q/\varepsilon_0$.

Solutions to Selected End-of-Chapter Problems

15. Picture the Problem: Three charges are arranged as indicated in the figure and exert electrostatic forces on each other.

$q_1 = +q$ $q_2 = -2.0\,q$ $q_3 = +3.0\,q$

d d

Strategy: Let the x-axis be along the line of the three charges with the positive direction pointing to the right. Use Coulomb's law (Equation 19-5) and the superposition of forces to find the net electrostatic force (magnitude and direction) on q_2. The force from q_1 will be attractive and to the left, and the force from q_3 will be attractive and to the right.

Solution: 1. (a) Write Coulomb's law using vector notation:

$$\vec{\mathbf{F}}_2 = \vec{\mathbf{F}}_{12} + \vec{\mathbf{F}}_{23} = -k\frac{q_1 q_2}{d^2}\hat{\mathbf{x}} + k\frac{q_1 q_3}{d^2}\hat{\mathbf{x}}$$

2. Substitute the charge magnitudes given in the figure

$$\vec{\mathbf{F}}_2 = \frac{k}{d^2}\left[-q_1 q_2 + q_1 q_3\right]\hat{\mathbf{x}} = \frac{k}{d^2}\left[-q(2.0q) + (2.0q)(3.0q)\right]\hat{\mathbf{x}}$$

$$= \frac{kq^2}{d^2}[4.0]\hat{\mathbf{x}} = [4.0]\frac{\left(8.99\times10^9\,\text{N}\cdot\text{m}^2/\text{C}^2\right)\left(12\times10^{-6}\,\text{C}\right)^2}{(0.16\,\text{m})^2}\hat{\mathbf{x}}$$

$$= \underline{\underline{(200\,\text{N})\hat{\mathbf{x}}}}$$

3. The net electrostatic force on q_2 is 200 N = $\boxed{0.20\text{ kN towards } q_3}$

4. (b) If the distance, d, were tripled, $\boxed{\text{the magnitude would be cut to a ninth}}$ and the direction would be unchanged.

Insight: The force is towards the right because q_3 is larger in magnitude and the distances are the same. We wrote the answer as 0.20 kN to emphasize there are only two significant figures, which wouldn't be clear if we wrote 200 N.

35. Picture the Problem: An electric charge experiences an upward force due to an electric field.

Strategy: The negative charge experiences an upward electric force, so we can conclude the electric field points downward. Set the magnitude of the electric force (Equation 19-9) equal to the magnitude of the weight in order to find the magnitude of the electric field. Then use the known electric field to find the force and hence the acceleration of the object when its charge is doubled. Let upward be the positive y direction.

Solution: 1. (a) Set $\vec{\mathbf{F}}_E = mg$ and solve for $\vec{\mathbf{E}}$:

$$|q|E = mg \Rightarrow E = \frac{mg}{|q|} = \frac{(0.012\,\text{kg})\left(9.81\,\frac{\text{m}}{\text{s}^2}\right)}{3.6\times10^{-6}\,\text{C}} = 3.3\times10^4\,\text{N/C}$$

$$\vec{\mathbf{E}} = E(-\hat{\mathbf{y}}) = \boxed{\left(-3.3\times10^4\,\text{N/C}\right)\hat{\mathbf{y}}}$$

2. (b) Use Newton's Second Law to find a: $\sum F_y = F_E - F_g = ma \Rightarrow (2q)E - mg = ma$

$$\therefore\ 2q(mg/q) - mg = ma \Rightarrow mg = ma$$

$$\therefore\ \vec{a} = g\,\hat{y} = \boxed{(9.81\ \text{m/s}^2)\,\hat{y}}$$

Insight: A negative charge is always pulled in a direction *opposite* the electric field.

41. Picture the Problem: The electric lines of force for this charge configuration are depicted in the figure at right.

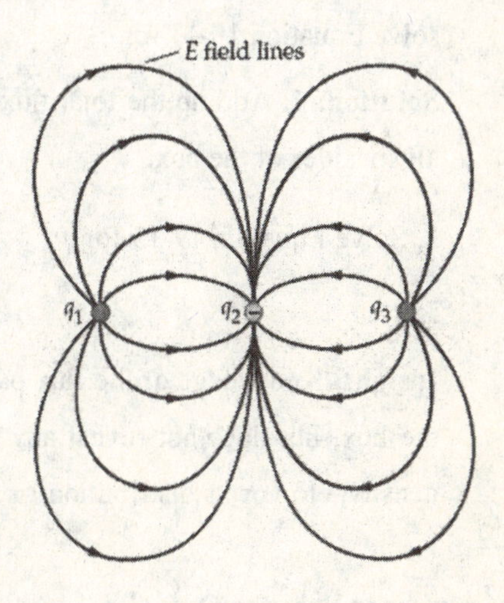

Strategy: Electric field lines diverge away from positive charges and converge towards negative charges. The number of lines of force entering or leaving a charge is proportional to the magnitude of the charge. Use these two rules and the accompanying diagram to determine the sign and magnitude of the charges q_1 and q_3.

Solution: 1. (a) The electric field lines begin at q_1 and q_3 and end at q_2. The rules of drawing electric field lines indicate that the charges q_1 and q_3 must be $\boxed{\text{positive}}$.

2. (b) The charge q_1 has 8 lines leaving it but q_2 has 16 lines entering it. Since 8 is half of 16, and since the number of lines entering or leaving a charge is proportional to the magnitude of the charge, the magnitude of q_1 is one-half of q_2, or $\boxed{5.00\ \mu\text{C}}$.

3. (c) By the reasoning of part **(b)**, the magnitude of q_3 is $\boxed{5.00\ \mu\text{C}}$

Insight: When drawn correctly, an electric field diagram can reveal a wealth of information.

43. Picture the Problem: Two charges, $+q$ and $-q$, are arranged as indicated in the figure at right.

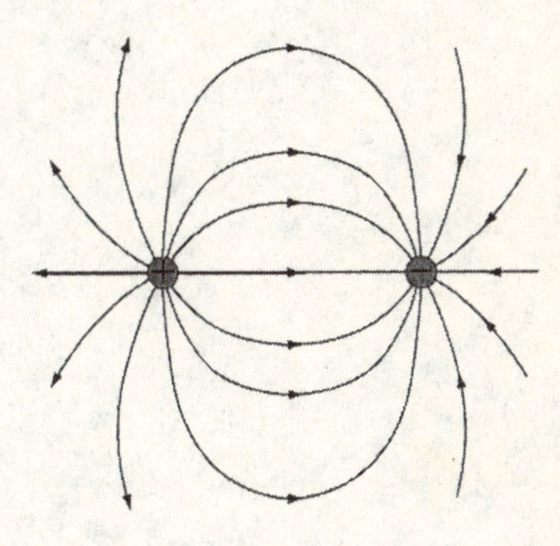

Strategy: Electric field lines diverge away from positive charges and converge towards negative charges. The number of lines of force entering or leaving a charge is proportional to the magnitude of the charge.

Solution: The lines of force are sketched at right.

Insight: When drawn correctly an electric field diagram can reveal a wealth of information.

53. Picture the Problem: A known amount of electric flux passes through each side of a rectangular box.

Strategy: Apply Gauss's law to the box, realizing that the total flux passing through all six sides of the box is proportional to the amount of charge enclosed by the box. Add up the total flux and then solve Equation 19-13 for q.

Solution: 1. Add up the total flux through all six sides of the box:

$$\Phi = \left(150.0 + 250.0 - 350.0 + 175.0 - 100.0 + 450.0\ \tfrac{\text{N}\cdot\text{m}^2}{\text{C}}\right) = \underline{\underline{575.0\ \tfrac{\text{N}\cdot\text{m}^2}{\text{C}}}}$$

2. Solve Equation 19-13 for q:

$$q = \varepsilon_0 \Phi = \left(8.85\times10^{-12}\ \frac{\text{C}^2}{\text{N}\cdot\text{m}^2}\right)\left(575.0\ \tfrac{\text{N}\cdot\text{m}^2}{\text{C}}\right) = \boxed{5.09\times10^{-9}\ \text{C}} = 5.09\ \text{pC}$$

Insight: Knowledge of the flux passing through each surface can tell us the amount of charge inside the box, but does not reveal any information about its character (point charges or uniform charge density, etc.) or its distribution.

Answers to Practice Quiz

1. (e) **2.** (d) **3.** (b) **4.** (d) **5.** (d) **6.** (c) **7.** (e) **8.** (a) **9.** (e) **10.** (a)

CHAPTER 20
ELECTRIC POTENTIAL and ELECTRIC POTENTIAL ENERGY

Chapter Objectives

After studying this chapter, you should

1. know the difference between electric potential and electric potential energy.
2. understand how both the electric potential and electric potential energy relate to the electric field.
3. be able to apply the conservation of energy to charged particles moving in electric fields.
4. be able to determine the electric potential of a configuration of point charges.
5. understand the relationship between equipotential surfaces and electric fields.
6. be able to calculate the capacitance of, electric field of, and energy stored in a parallel-plate capacitor with and without a dielectric.

Warm-Ups

1. Suppose you are given three capacitors consisting of aluminum disks configured as in the figure. You put equal amounts of positive charge on the top disks of each of the three sets and corresponding amounts of negative charge on the bottom disks. If you were to measure the electric fields between the disks, which set would give you the highest value? Which one the lowest? How about the voltages?

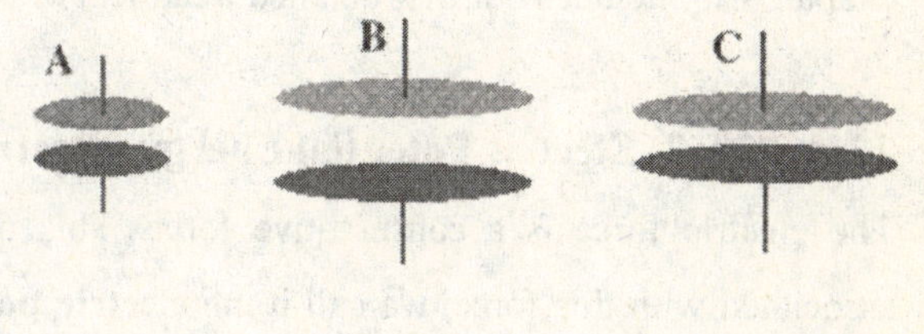

2. Electronic flashes in cameras flash when a capacitor is discharged through a material that produces light energy when electric charge passes through it. This energy comes from a battery inside the flash unit. Estimate the capacitance of the capacitor used to store the energy. (Hint: Assume that the flash gives you the same amount of light as a 500-watt bulb that is turned on for a 60th of a second).
3. A fresh "D cell" battery can provide about 5000 J of electrical energy before it must be discarded. Estimate the number of coulombs that must pass through it during its lifetime.

4. The voltage across a capacitor is given by the formula $V = Q/C$, where Q is usually called "the charge on the capacitor." Where is this charge in a capacitor? Does the capacitor really have a net charge? If two capacitors are connected together, is the charge on one the same as on the other?
5. You and a close friend stand facing each other. You are as close as you can get without actually touching. If a wire is attached to each of you, you can act as the two conductors in a capacitor. Estimate the capacitance of this "human capacitor."
6. Contour maps like the one to the right show equally spaced lines of constant elevation at Earth's surface. Explain in what sense they are "equally spaced." Also, explain in what sense these lines could be called gravitational equipotentials

7. Estimate the total charge that passes through a light bulb in one second.

Chapter Review

In this chapter, we study the potential energy associated with the electric force and the electric field. This chapter also includes our first detailed treatment of an important electrical device called a capacitor.

20–1 – 20–2 Electric Potential Energy, Electric Potential, and Energy Conservation

The electric force is a conservative force. This means that it is useful to define a potential energy associated with this force; we call it the **electric potential energy**. As always, only changes in potential energy are measurable, and so we define only the change in potential energy, ΔU. The change in electric potential energy is defined as the negative of the work done by the electric force

$$\Delta U = -W_E$$

Because work done depends on the force applied, and the electric force, in turn, depends on the charge on which it is applied, the change in potential energy must depend on the charge. Recall that the concept of the electric field (force per unit charge) gives us a way to handle information about the electric force without reference to any specific charge; in a similar way we can define a quantity, the **electric potential**, to handle information about the energy without reference to a specific charge. The electric potential difference, ΔV, is defined as

$$\Delta V = \frac{\Delta U}{q_0}$$

where q_0 is a test charge that moves from one point to another. The SI unit of electric potential is J/C and is called a **volt** (V): 1V = 1 J/C. Often, instead of electric potential, the quantity V is called the *voltage*. The above definition also implies that $\Delta U = q\,\Delta V$; a commonly used unit of energy based on this relation is the **electron-volt** (eV). An eV is the energy change experienced by an electron or proton when it accelerates through a potential difference of 1 V:

$$1\text{ eV} = \left(1.60\times10^{-19}\text{ C}\right)\left(1\text{ V}\right) = 1.60\times10^{-19}\text{ J}$$

The above definitions imply a direct relationship between ΔV and the electric field, E. It works out that the electric field can be determined from the rate at which the electric potential changes with position,

$$E = -\frac{\Delta V}{\Delta s}$$

where Δs represents the displacement from one position to another. The minus sign indicates that the electric field points in the direction of decreasing electric potential. This means that positive charges accelerate in the direction of decreasing electric potential and negative charges accelerate in the direction of increasing electric potential. The preceding relation also shows that an alternative unit for electric field could be V/m instead of N/C. In fact, both units of E are in common use — 1 N/C = 1 V/m.

That we can define an electric potential energy also means that we can extend the conservation of mechanical energy to the motion of charged particles in electric fields; therefore, we can write that the sum of the kinetic and electric potential energies must be constant: $K_i + U_i = K_f + U_f$. At a given location $U = qV$, therefore, the above result can also be written in terms of the electric potential as

$$K_i + qV_i = K_f + qV_f$$

Example 20–1 Motion in a Constant Electric Field A particle of charge 22.4 mC and mass 57.2 mg is initially held at rest in a constant electric field of 133 V/m. If the particle is released and accelerates through a distance of 14.9 m, **(a)** determine its change in potential energy, **(b)** determine the change in electric potential it experiences, and **(c)** use energy conservation to determine its final speed.

Picture the Problem The diagram shows the electric field lines and the charged particle.

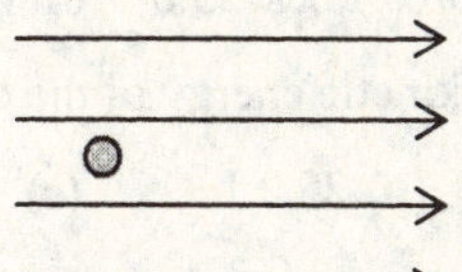

Strategy The given information allows us to calculate the work on the particle. By knowing this, we can use the definitions of electric potential energy and electric potential to find the answers.

Solution

Part (a)

1. Since the force and displacement are in the same direction, the work done by the electric force is: $W_E = F_E d = qEd$

2. By definition, the change in potential energy must be: $\Delta U = -W_E = -qEd = -(22.4\times10^{-3}\text{ C})(133\ \tfrac{\text{N}}{\text{C}})(14.9\text{ m}) = -44.4\text{ J}$

Part (b)

3. From the definition of electric potential, we obtain ΔV: $\Delta V = \dfrac{\Delta U}{q} = \dfrac{-44.39\text{ J}}{0.0224\text{ C}} = -1.98\text{ kV}$

Part (c)

4. According to energy conservation: $\Delta K = -\Delta U \Rightarrow K_f - 0 = -\Delta U \therefore \frac{1}{2}mv_f^2 = -\Delta U$

5. Solving for the final speed gives: $v_f = \sqrt{\dfrac{-2(\Delta U)}{m}} = \sqrt{-\dfrac{2(-44.39\text{ J})}{57.2\times10^{-6}\text{ kg}}} = 1.25\times10^3\text{ m/s}$

Insight For convenience, I switched the unit of electric field from V/m to N/C; remember V/m = N/C.

Practice Quiz

1. An electric force does 25.0 J of work in accelerating a particle of charge –15.0 μC. What is the change in the electric potential energy of the charge?

   (a) 15.0 J **(b)** 25.0 J **(c)** –15.0 J **(d)** –25.0 J **(e)** 0 J

2. An electric force does 25.0 J of work in accelerating a particle of charge –15.0 μC. What change in electric potential does the charge experience?

   (a) 1.67 MV **(b)** –1.67 MV **(c)** –25.0 MV **(d)** 25.0 MV **(e)** 0 MV

3. An electric force does 25.0 J of work in accelerating a particle of charge –15.0 μC. What is the change in the kinetic energy of the charge?

   (a) 15.0 J **(b)** 25.0 J **(c)** –15.0 J **(d)** –25.0 J **(e)** 0 J

20–3 The Electric Potential of Point Charges

A detailed treatment of point charges using the definitions of the previous section reveals that for a test charge moving from point B to A in the presence of a charge, q, fixed at the origin, the change in electric potential energy is

$$\Delta U = kqq_0\left(\frac{1}{r_A}-\frac{1}{r_B}\right)$$

In order to assign a value of potential energy to a given charge configuration, a reference configuration must be chosen at which we set $U = 0$. For point charges, this reference is typically chosen to be infinitely far away from the charge q. With this choice of reference understood, we can say that the value of the potential energy of two point charges, q and q_0, separated by a distance, r, is

$$U = \frac{kqq_0}{r}$$

Applying the definition of the electric potential to the above results for the potential energy suggests that the value of the electric potential a distance r away from a point charge, q, can be taken as

$$V = \frac{kq}{r}$$

This result is for the electric potential of a single point charge. For a system of two or more point charges, we can use the fact that the electric potential obeys a principle of superposition appropriate for a scalar quantity:

The electric potential due to two or more point charges equals the algebraic sum of the potentials due to each point charge individually.

The electric potential energy of a system of three or more point charges obeys a similar superposition principle:

The electric potential energy of a system of three or more point charges equals the algebraic sum of the potential energy of each pair of charges.

Example 20–2 Potential and Potential Energy A configuration of three charged particles is such that charge q_1 = 5.3 nC has (x, y) coordinates of (1.5 m, 0.0 m), charge q_2 = 9.2 nC is located at (0.0 m, 2.0 m), and charge q_3 = –2.4 nC is located at the origin. **(a)** Determine the electric potential at position P = (1.5 m, 2.0 m). **(b)** Determine the change in potential energy if a 7.4-nC charge is placed at P from infinitely far away.

Picture the Problem The diagram shows the charge configuration of q_1, q_2, and q_3. The point P is indicated by the open circle.

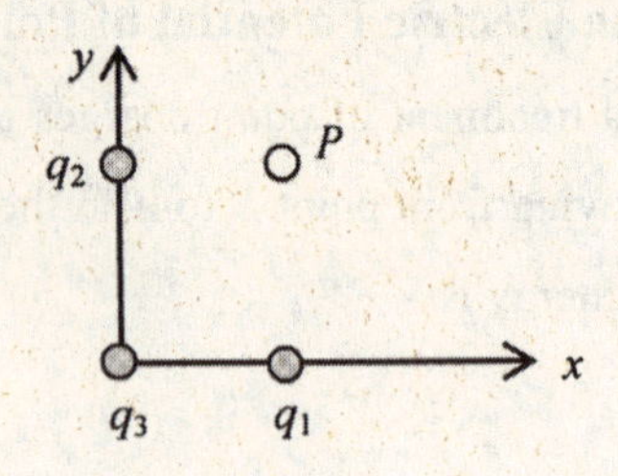

Strategy We can solve this problem by using the result for the potential energy due to a point charge together with the principle of superposition.

Solution

Part (a)

1. The potential at P is the sum of the potentials due to each charge: $V_P = V_1 + V_2 + V_3 = k\left(\frac{q_1}{r_1} + \frac{q_2}{r_2} + \frac{q_3}{r_3}\right)$

2. The distances can be determined from the coordinates: $r_1 = 2.0 \text{ m},\quad r_2 = 1.5 \text{ m}$

   $r_3 = \sqrt{(1.5 \text{ m})^2 + (2.0 \text{ m})^2} = 2.5 \text{ m}$

3. The potential at P then is: $V_P = \left(8.99 \times 10^9 \frac{\text{N} \cdot \text{m}^2}{\text{C}^2}\right)\left(\frac{5.3 \text{ nC}}{2.0 \text{ m}} + \frac{9.2 \text{ nC}}{1.5 \text{ m}} - \frac{2.4 \text{ nC}}{2.5 \text{ m}}\right) = 70 \text{ V}$

Part (b)

4. We can use the above result to get the change in potential energy if we bring in the fourth charge: $\Delta U = q(\Delta V) = (7.4 \text{ nC})(70.3 \text{ V}) = 5.2 \times 10^{-7} \text{ J}$

Insight In part (b), most of the information for getting ΔU was already determined when we calculated V at P. The result for V_p equaled ΔV for the fourth charge because at infinity, $V = 0$.

Practice Quiz

4. Two point charges, $q_1 = 23$ μC and $q_2 = -75$ μC, are moved such that the distance between them changes from 17.3 m to 4.77 m. What is the change in the potential energy of this system of charges?

   (a) –2.4 J **(b)** 2.4 J **(c)** – 3.3 J **(d)** – 0.90 J **(e)** 0.90 J

5. Two point charges, $q_1 = 23$ μC and $q_2 = -75$ μC, are separated by a distance of 13 m. What is the electric potential at the midpoint between them?

   (a) 0 V **(b)** –100 kV **(c)** –32 kV **(d)** – 36 kV **(e)** –72 kV

20–4 Equipotential Surfaces and the Electric Field

Recall that to get a visual feel for the electric field of a configuration of charges, we can draw electric field lines. In a similar way, we can obtain a visual representation of the electric potential due to a configuration of charges by locating the surfaces on which the potential will have the same value. These surfaces are called **equipotential surfaces**. In two-dimensional drawings, these surfaces are often represented by simple lines and are often called *equipotential lines*, or just *equipotentials*.

Because, by definition, there is no change in electric potential when a charge moves along an equipotential surface, we know that no work is done on this particle by the electric force; therefore, the electric field must not have any component along an equipotential surface. This result leads to the useful fact that

the electric field is always perpendicular to the equipotential surfaces.

That charged particles on the surfaces of conductors configure themselves such that there is no net electric field along the surface, indicates that conducting surfaces are also equipotential surfaces. As a result of this fact, charge becomes more concentrated near sharp edges on a conductor than near blunt ends. The above discussion implies that when a conducting surface is in an electric field, in static equilibrium, the electric field must be perpendicular to the surface.

Exercise 20–3 Estimating the Electric Field Two equipotential surfaces of 28.2 V and 29.9 V are separated by 3.36 mm. Estimate the magnitude of the average electric field between the surfaces.

Solution We are given the following information:

Given: $V_1 = 28.2$ V, $V_2 = 29.9$ V, $\Delta x = 3.36$ mm; **Find:** E

Here we can make direct use of the relationship between the electric field and the electric potential. However, because we only want the magnitude of the electric field, we can ignore minus signs:

$$E = \frac{\Delta V}{\Delta x} = \frac{V_2 - V_1}{\Delta x} = \frac{29.9 \text{ V} - 28.2 \text{ V}}{3.36 \times 10^{-3} \text{ m}} = 506 \text{ V/m}$$

Practice Quiz

6. A particle of charge –15.0 μC is moved a distance of 2.00 m along an equipotential surface of value –24.0 V. What is the change in the potential energy of the charge?

   (a) 3.60×10^{-4} J **(b)** 0 J **(c)** 7.20×10^{-4} J **(d)** 12.0 J **(e)** None of the above.

20–5 – 20–6 Capacitors, Dielectrics, and Electrical Energy Storage

A **capacitor** — an important device used in many electrical applications — consists of two conductors separated by some distance. The conductors that make up a capacitor are generically called *plates*. Capacitors store charge and electrical energy when a potential difference, V, is applied across the plates. Equal and opposite amounts of charge is stored on each plate. The magnitude of charge, Q, that can be stored is characterized by its **capacitance,** C, which is defined as

$$C = \frac{Q}{V}$$

The charge, Q, and the potential difference, V, are usually related to each other in such a way that the capacitance typically does not depend on either. Rather, the capacitance is determined by the size, shape, and distance between the conductors. As can be seen from the preceding expression, the SI unit of capacitance is C/V, which is called a farad (F): 1 F = 1 C/V. In practice, 1 F is a large amount of capacitance; therefore, we usually work with capacitances in the range between picofarad (1 pF = 10^{-12} F) and microfarad (1 μF = 10^{-6} F).

A prototype for studying capacitance is the parallel-plate capacitor, which consists of two flat, parallel plates, each of area, A, separated by a distance, d. Two useful facts about parallel-plate capacitors are (a) the electric field between the plates is uniform (except near the edges), and its magnitude is given by

$$E = \frac{Q}{\varepsilon_0 A}$$

and (b) the capacitance of a parallel-plate capacitor is given by

$$C = \frac{\varepsilon_0 A}{d}$$

These results assume that there is vacuum between the plates of the capacitor. A common practice that increases the capacitance of a capacitor is to insert an insulating material, called a *dielectric*, between the plates. When a dielectric is present in a capacitor, polarization of the molecules that make up the dielectric gives rise to a reduced electric field between the plates:

$$E = \frac{E_0}{\kappa}$$

where E_0 is the electric field in vacuum, and κ is a dimensionless quantity greater than 1, called the **dielectric constant**, that depends on the material being used. Similarly, in the presence of a dielectric, the potential difference between the plates is reduced from the value, V_0, in vacuum by a factor of κ:

$$V = \frac{V_0}{\kappa}$$

These results imply that the capacitance is increased in the presence of a dielectric by a factor of κ:

$$C = \kappa C_0$$

For the particular case of a parallel-plate capacitor containing a dielectric we have

$$C = \frac{\kappa \varepsilon_0 A}{d}$$

It takes energy to store an amount of charge Q on an initially uncharged capacitor; this energy is stored as electric potential energy in the capacitor. The amount of energy U stored in a capacitor that holds a charge Q for a potential difference V across its plates can be determined by three equivalent expressions:

$$U = \tfrac{1}{2}QV = \tfrac{1}{2}CV^2 = \frac{Q^2}{2C}$$

This energy is stored in the electric field that exists between the plates of the capacitor. The relationship between the stored energy and the electric field is most conveniently written in terms of the energy density (energy per volume) of the electric field:

$$\text{energy density} = \tfrac{1}{2}\varepsilon_0 E^2$$

Example 20–4 A Parallel-plate Capacitor A certain parallel-plate capacitor has plates of area 13.7 cm^2 that are separated by a distance of 1.55 cm. If a material with a dielectric constant of 2.0 exists between the plates, and a potential difference of 2.50 V is applied across them, **(a)** how much charge will the plates hold, and **(b)** how much energy is stored in the capacitor?

Picture the Problem The sketch shows a side view of the plates of a parallel-plate capacitor containing a dielectric.

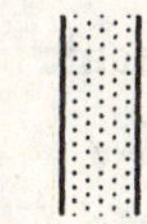

Strategy Because it is a parallel-plate capacitor we know how to calculate the capacitance; because we are given the potential difference, we can determine the charge and energy stored.

Solution

Part (a)

1. The capacitance is given by: $C = \frac{\kappa \varepsilon_0 A}{d}$

2. Therefore, the charge stored is:

$$Q = VC = \frac{V\kappa\varepsilon_0 A}{d}$$

$$= \frac{(2.50\text{ V})2.0\left(8.85\times10^{-12}\ \frac{\text{C}^2}{\text{N}\cdot\text{m}^2}\right)\left(13.7\times10^{-4}\text{ m}^2\right)}{1.55\times10^{-2}\text{ m}}$$

$$= 3.91\text{ pC}$$

Part (b)

3. The energy stored is:

$$U = \tfrac{1}{2}QV = \tfrac{1}{2}\left(3.911\times10^{-12}\text{ C}\right)(2.50\text{ V}) = 4.9\text{ pJ}$$

Insight We did not need a numerical calculation of the capacitance, so to eliminate the possibility of round-off error, we did not do one.

Practice Quiz

7. A 2.0-μF capacitor is connected across a 4.0-V battery. How much charge is stored on the negative plate?

   (a) –8.0 μC **(b)** –0.50 μC **(c)** –6.0 μC **(d)** –2.0 μC **(e)** –1.0 μC

8. Which of the following statements is *not* correct for an isolated charged capacitor?

   (a) Inserting a dielectric into a previously empty capacitor decreases the potential difference across it.

   (b) Inserting a dielectric into a previously empty capacitor decreases the electric field across it.

   (c) Inserting a dielectric into a previously empty capacitor increases the capacitance.

   (d) Inserting a dielectric into a previously empty capacitor allows it to store more charge.

   (e) None of the above.

9. A certain parallel-plate capacitor stores an amount of energy U between its plates. If the potential difference across it is doubled, the energy stored equals

   (a) U. **(b)** $2U$. **(c)** $4U$. **(d)** $U/2$. **(e)** $U/4$.

10. A certain parallel-plate capacitor stores an amount of energy U between its plates. If the charge stored on it is doubled, the energy stored equals

    (a) U. **(b)** $2U$. **(c)** $4U$. **(d)** $U/2$. **(e)** $U/4$.

Reference Tools and Resources

I. Key Terms and Phrases

electric potential energy the potential energy associated with the electric force

electric potential the electric potential energy per unit charge

volt the SI unit of electric potential

electron-volt a unit of energy equal to 1.60×10^{-19} J

equipotential surface a surface on which every point has the same value of electric potential

capacitor an electrical device, used to store charge and energy, that consists of two conductors separated by a finite distance

capacitance a measure of a capacitor's ability to store charge

dielectric an insulating material often used between the plates of a capacitor

energy density energy per unit volume

II. Important Equations

Name/Topic	Equation	Explanation
electric potential energy	$\Delta U = -W_E$	The definition of electric potential energy
electric potential difference	$\Delta V = \frac{\Delta U}{q_0}$	The definition of electric potential
electric potential	$V = \frac{kq}{r}$	The electric potential of a point charge q assuming that $V = 0$ at $r = \infty$
capacitors	$C = \frac{Q}{V}$	The definition of capacitance
	$C = \frac{\kappa\varepsilon_0 A}{d}$	The capacitance of a parallel-plate capacitor with a dielectric
	$U = \frac{1}{2}QV = \frac{1}{2}CV^2 = \frac{Q^2}{2C}$	The energy stored in a capacitor

III. Know Your Units

Quantity	Dimension	SI Unit
electric potential (V)	$[M][L^2][T^{-2}][C^{-1}]$	V
capacitance (C)	$[C^2][T^2][M^{-1}][L^{-2}]$	F

Puzzle

STUFFED CAPACITOR

Consider the figures. The left figure, is a plain parallel-plate capacitor (area = A, gap = d). The right figure is a system consisting of the same capacitor with a metal plate inserted into the gap. Assume the thickness of the plate is $d/2$. How will the new capacitance compare with the old? Answer this question in words, not equations, briefly explaining how you obtained your answer.

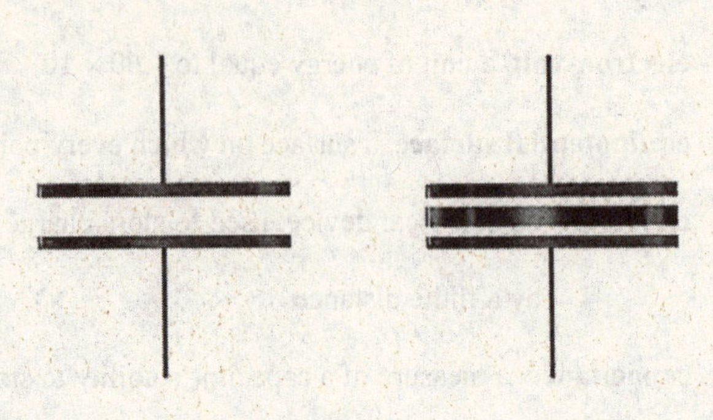

Answers to Selected Conceptual Questions and Exercises

Conceptual Questions

2. The two like charges, if released, will move away from one another to infinite separation, converting the positive electric potential energy into kinetic energy. The two unlike charges, however, attract one another – if their separation is to be increased, a positive work must be done. In fact, the minimum amount of work that must be done to create an infinite separation between the charges is equal to the magnitude of the original negative electric potential energy.

8. Not necessarily. The electric potential energy of the two charges would be the same only if the charges are also equal. In general, the electric potential energy is the charge times the electric potential. Therefore, just having equal potentials does not guarantee equal potential energy.

12. A capacitor stores charge of opposite sign in two different locations – though, of course, the net charge is zero. We can think of a capacitor, then, as storing a "charge separation", along with the energy required to cause the charge separation in the first place.

Conceptual Exercises

6. **(a)** Point B is closer to q_2 than to q_1 by a factor of the square root of 2. Therefore, if the electric potential is to be zero at point 2, it is necessary that q_2 be negative, and have a magnitude that is less than the magnitude of q_1 by a factor of $\sqrt{2}$. We conclude, then, that $q_2 = -Q/\sqrt{2}$. **(b)** Notice that point A is closer to the positive charge, +Q than to the negative charge, $-Q/\sqrt{2}$, and that the negative charge has a smaller magnitude. It follows that the electric potential at point A is positive.

10. **(a)** Note that electric potential is proportional to the charge divided by the distance from the charge. It follows, then, that the electric potential is zero in this case at a point on the x axis that is twice as far from $+2q$ as it is from $-q$. The point between the two charges where this condition is satisfied is $x = -0.5$ m, which is 2.0 m from $+2q$ and 1.0 m from $-q$. **(b)** Again, the condition to be satisfied is that the point in question must be twice as far from $+2q$ (at $x = 1.5$ m) as it is from $-q$ (at $x = -1.5$ m). This can occur only in region 3. **(c)** The point referred to in part (b) is $x = -4.5$ m, which is 3.0 m from $-q$ and 6.0 m from $+2q$.

18. **(a)** The electric field between the plates remains the same, because both the potential difference and the plate separation are the same. It follows that the rate of change of potential with distance (the magnitude of the electric field) is unchanged. **(b)** There is more charge on the plates now, because more charge is required to produce the same electric field in the presence of the polarizing effect of the dielectric. **(c)** The capacitor stores more charge for the same potential difference; therefore, its capacitance is increased. **(d)** Recall that the energy stored in a capacitor with the potential difference V between its plates is $U = \frac{1}{2}QV = \frac{1}{2}CV^2$. We know that the dielectric increases the charge [part (b)] and also increases the capacitance [part (c)]. It follows that the stored energy is increased as well.

Solutions to Selected End-of-Chapter Problems

11. **Picture the Problem**: A uniform electric field $\vec{\mathbf{E}} = (-1200\text{ N/C})\hat{\mathbf{x}}$ creates a change in the electric potential along the $\hat{\mathbf{x}}$ direction.

Strategy: The change in electric potential is given by Equation 20-4, $\Delta V = -E\Delta s$, where Δs is equal to Δx because the field points in the $\hat{\mathbf{x}}$ direction. There is no change in the electric potential along the $\hat{\mathbf{y}}$ and $\hat{\mathbf{z}}$ directions. The change in electric potential between any two points is independent of the path taken between those two points.

Solution: 1. (a) $\Delta x = 0$ between points A and B: $\Delta V = V_B - V_A = -E\Delta x = \boxed{0}$

2. (b) Solve Equation 20-4 for ΔV between points B and C: $\Delta V = V_B - V_C = -E\Delta x = -E(x_B - x_C) = -(-1200\text{ N/C})(-0.040\text{ m}) = \boxed{-48\text{ V}}$

3. (c) Repeat for points C and A: $\Delta V = V_C - V_A = -E\Delta x = -E(x_C - x_A) = -(-1200\text{ N/C})(0.040\text{ m}) = \boxed{48\text{ V}}$

4. (d) $\boxed{\text{No}}$, we cannot determine the value of the electric potential at point A because the location of zero potential has not been established. We could always choose $V_A = 0$ if we wished.

Insight: Note that the potential changes only along the direction parallel to the field, and is higher at point C than at either point A or point B. This is consistent with statement, "The electric field points downhill on the potential surface," as described by figure 20-3 and the accompanying text.

19. Picture the Problem: A charged particle accelerates in an electric field.

Strategy: The gain in kinetic energy of the particle equals the loss in potential energy, which equals the magnitude of the field times the distance parallel to the field, which is oriented in the $\hat{\mathbf{x}}$ direction. The magnitude of the electric field is given in problem 11 as 1200 N/C.

Solution: 1. (a) The particle has a positive charge, so it will move in the direction of the electric field, which is the $\boxed{\text{negative } x\text{-direction}}$.

2. (b) Set $\Delta K = -\Delta U$ and solve for v: $\frac{1}{2}mv^2 = -q\Delta V = -qE\Delta x$

$$v = \sqrt{\frac{-2qE\Delta x}{m}} = \sqrt{\frac{-2\left(0.045\times10^{-6}\text{ C}\right)\left(1200\,\frac{\text{N}}{\text{C}}\right)\left(-0.050\text{ m}\right)}{0.0035\text{ kg}}} = \boxed{3.9\,\frac{\text{cm}}{\text{s}}}$$

3. (c) The increase in speed over the second 5.0 cm will be $\boxed{\text{less than}}$ its increase in speed in the first 5.0 cm because v is proportional to the square root of the distance traveled from the rest position.

Insight: Verify for yourself that the particle gains 1.6 cm/s of speed over the second 5.0 cm for a final speed of 5.5 cm/s.

31. Picture the Problem: Three charges are arranged at the corners of a rectangle as indicated in the diagram at right.

Strategy: The work required to move the 2.7 μC charge to infinity is equal to the change in its electric potential energy due to the other two charges. Sum the electric potential energies due to the other two charges to find the electric potential energy at the corner of the rectangle. At infinity the electric potential energies are zero. Use these facts to find the work required to move the 2.7 μC charge to infinity, and then repeat the procedure to find the work required to move the −6.1 μC charge.

Solution: 1. (a) Find $\Delta U = U_f - U_i = 0 - \left(U_1 + U_2\right)$:

$$W = \Delta U = \left[0 - \left(\frac{kq_2q_1}{r_{21}} + \frac{kq_2q_3}{r_{23}}\right)\right] = -kq_2\left(\frac{q_1}{r_{21}} + \frac{q_3}{r_{23}}\right)$$

$$= -\left(8.99\times10^9\,\frac{\text{N}\cdot\text{m}^2}{\text{C}^2}\right)\left(2.7\times10^{-6}\text{ C}\right)\left[\frac{-6.1\times10^{-6}\text{ C}}{0.25\text{ m}} + \frac{-3.3\times10^{-6}\text{ C}}{\sqrt{(0.25\text{ m})^2 + (0.16\text{ m})^2}}\right]$$

$$W = \boxed{0.86\text{ J}}$$

2. (b) The $-6.1\ \mu\text{C}$ charge is repelled by the $-3.3\ \mu\text{C}$ charge more than it is attracted by the $2.7\ \mu\text{C}$ charge. The work required will be negative, which is $\boxed{\text{less than}}$ the work required in part (a).

3. (c) Repeat step 1 for the $-6.1\ \mu\text{C}$ charge:

$$W = \Delta U = \left[0 - \left(\frac{kq_1q_2}{r_{12}} + \frac{kq_1q_3}{r_{13}}\right)\right] = -kq_1\left(\frac{q_2}{r_{12}} + \frac{q_3}{r_{13}}\right)$$

$$= -\left(8.99\times10^9\ \tfrac{\text{N}\cdot\text{m}^2}{\text{C}^2}\right)\left(-6.1\times10^{-6}\ \text{C}\right)\left[\frac{2.7\times10^{-6}\ \text{C}}{0.25\ \text{m}} + \frac{-3.3\times10^{-6}\ \text{C}}{0.16\ \text{m}}\right]$$

$$W = \boxed{-0.54\ \text{J}}$$

Insight: Note that the work done by the electric field is $W = -\Delta U$ (Equation 20-1) but the work you must do to move the charge against the field is $W = \Delta U$. The negative work of part (c) means the charge would fly off on its own.

39. Picture the Problem: The electric field points downhill on the potential surface and its magnitude is the slope of that surface.

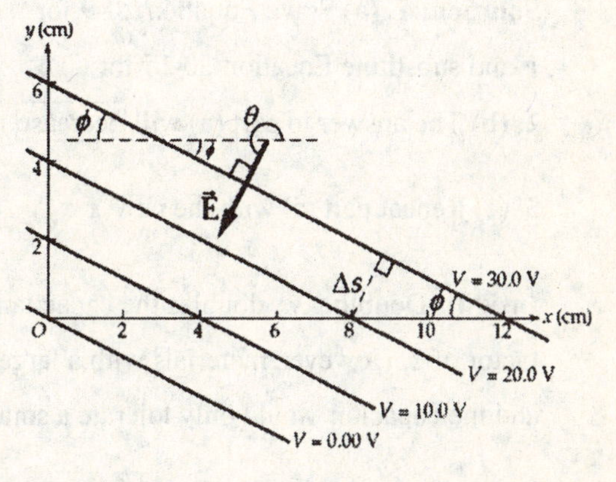

Strategy: We can see by examining the figure that the potential changes by 10.0 V along a distance, Δs, in the direction perpendicular to the equipotential lines. Find Δs using geometry and apply Equation 20-4 to find the magnitude of $\vec{\mathbf{E}}$. Since $\vec{\mathbf{E}}$ must be perpendicular to the equipotential lines and point towards decreasing potential, we can use trigonometry to determine its direction. Finally, the shortest distance a charge can move to undergo a change in potential of 5.00 V is half of Δs.

Solution: 1. (a) Find ϕ from the right triangle formed by the axes and the V = 30.0 V equipotential:

$$\phi = \tan^{-1}\left(\frac{6.00\ \text{cm}}{12.0\ \text{cm}}\right) = \underline{\underline{26.6^\circ}}$$

2. Find Δs from the smaller triangle at the right side of the figure:

$$\sin\phi = \frac{\Delta s}{4.00\ \text{cm}} \Rightarrow \Delta s = (4.00\ \text{cm})\sin 26.6^\circ = \underline{\underline{1.79\ \text{cm}}}$$

3. Find E from Equation 20-4:

$$E = \left|-\frac{\Delta V}{\Delta s}\right| = \frac{10.0\ \text{V}}{0.0179\ \text{m}} = \boxed{559\ \text{V/m}}$$

4. Find the direction of $\vec{\mathbf{E}}$ from the triangle it forms with the 30.0 V line:

$$\theta = 180^\circ + (90^\circ - \phi) = 180^\circ + (90^\circ - 26.6^\circ) = \boxed{243^\circ}$$

5. (b) A 5.00 V change occurs over a distance of $\frac{1}{2}\Delta s = \frac{1}{2}(17.9\text{ mm}) = \boxed{8.95\text{ mm}}$:

Insight: Not only does $\vec{\mathbf{E}}$ point "downhill" on the potential surface, but its magnitude equals the *steepest* slope at that location. That is why $\Delta s = \Delta V/E$ yields the minimum distance a charge can move to undergo a certain ΔV. See Problem 74 for an alternative method of solving this problem.

45. Picture the Problem: A capacitor is formed with two parallel plates separated by a dielectric layer.

Strategy: Equation 20-15 gives the relationship between the capacitance of a parallel-plate capacitor, the plate area *A*, the plate separation *d*, and the dielectric constant κ. Substitute this expression for the capacitance into Equation 20-9, $C = Q/V$ and solve for *V*.

Solution: 1. (a) Solve Equation 20-9 for *V* and substitute Equation 20-15 for *C*:

$$V = \frac{Q}{C} = \frac{Qd}{\kappa\varepsilon_0 A} = \frac{(4.7\times10^{-6}\text{ C})(0.88\times10^{-3}\text{ m})}{(2.0)(8.85\times10^{-12}\frac{\text{C}^2}{\text{N}\cdot\text{m}^2})(0.012\text{ m}^2)} = \boxed{19\text{ kV}}$$

2. (b) The answer to part (a) will $\boxed{\text{decrease}}$ if κ is increased because *V* is inversely proportional to κ.

3. (c) Repeat part (a) with the new κ:

$$V = \frac{Qd}{\kappa\varepsilon_0 A} = \frac{(4.7\times10^{-6}\text{ C})(0.88\times10^{-3}\text{ m})}{(4.0)(8.85\times10^{-12}\frac{\text{C}^2}{\text{N}\cdot\text{m}^2})(0.012\text{ m}^2)} = \boxed{9.7\text{ kV}}$$

Insight: Doubling κ doubles the capacitance and reduces the required voltage to store charge *Q* by a factor of 2. However, materials with a large κ also tend to have small dielectric strength (Table 20-2) and the capacitor would only tolerate a smaller potential difference across the plates.

59. Picture the Problem: A capacitor in a camera flash unit stores energy when it is charged up to a certain voltage.

Strategy: Solve $C = Q/V$ (Equation 20-9) for *Q* and then apply $U = \frac{1}{2}CV^2$ (Equation 20-17) to find the energy *U* stored in the capacitor.

Solution: 1. (a) Solve Equation 20-9 for *Q*: $Q = CV = (850\times10^{-6}\text{ F})(330\text{ V}) = \boxed{0.28\text{ C}}$

2. (b) Apply Equation 20-17 directly: $U = \frac{1}{2}CV^2 = \frac{1}{2}(850\times10^{-6}\text{ F})(330\text{ V})^2 = \boxed{46\text{ J}}$

Insight: The entire 46 J is deposited into the lamp in a very short time, producing a very bright flash. Special circuitry in the flash unit can build up the very large potential difference across the capacitor using low voltage, dry cell batteries.

Answers to Practice Quiz

1. (d) **2.** (a) **3.** (b) **4.** (a) **5.** (e) **6.** (b) **7.** (a) **8.** (e) **9.** (c) **10.** (c)

CHAPTER 21

ELECTRIC CURRENT and DIRECT-CURRENT CIRCUITS

Chapter Objectives

After studying this chapter, you should

1. understand the meaning of electric current.
2. be able to apply Ohm's law to basic circuits involving resistors.
3. know how to determine power consumption in electric circuits.
4. be able to correctly combine resistors and capacitors in series and parallel.
5. be able to apply Kirchhoff's rules to electric circuits.
6. understand the behavior of RC circuits.
7. know how ammeters and voltmeters should be properly connected into circuits.

Warm-Ups

1. In the circuit shown, the battery provides a current $I = 0.3$ A to the resistor. How much current is returned by the resistor to the negative terminal of the battery? How much current flows inside the battery from the negative to the positive terminal?

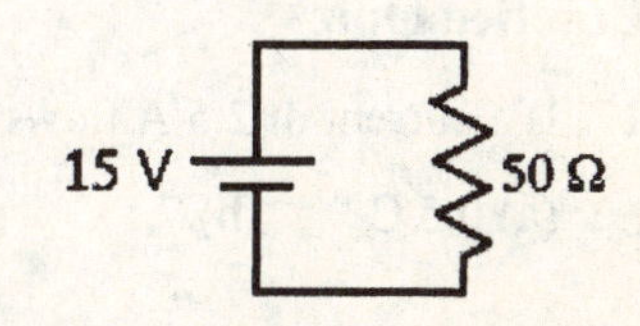

2. Estimate the resistance of a typical electrical cord in your home, say, one that attaches a lamp to the wall socket.
3. According to your textbook, the power dissipated by a resistor is given by $P = V^2/R$. However, your textbook also says that the power dissipated by a resistor is given by $P = I^2R$. Is the power really proportional to R or is it really proportional to $1/R$? Can both be correct?
4. Estimate the resistance of a typical light bulb. (Hint: Use "typical" values for the voltage and power.)

Chapter Review

In this chapter, we study direct current circuits. The important topics of Ohm's law and Kirchhoff's rules are covered. Capacitors as circuit elements are also introduced.

21–1 Electric Current

The flow of electric charge constitutes an **electric current,** I. The current is determined by the net magnitude of charge, ΔQ, that flows past a point in time Δt,

$$I = \frac{\Delta Q}{\Delta t}$$

The SI unit of current is called an ampere, A, 1 A = 1 C/s. For technical reasons, the ampere, not the coulomb, is considered to be the fundamental SI unit. From now on, we will take the fundamental dimension used for electricity to be [A].

In this chapter, we consider currents that flow through closed paths called **electric circuits**. Since the flow of current here will be in one direction around the circuit, these are called **direct current circuits**. A current flows in a circuit when energy is supplied by an **electric battery**. Batteries have two ends called *terminals* across which a potential difference exists. This potential difference is called the **electromotive force,** $\mathcal{E}$, or emf, of the battery. The direction of the flow of current in a circuit is always taken to be the direction in which a positive charge would move. This last statement is true even if the actual charge that flows is negative, as is the case in almost all electrical devices.

Practice Quiz

1. If a current of 2.5 A flows through a wire for 3.0 seconds, how much charge passes through the wire?
   (a) 0.5 C **(b)** 7.5 C **(c)** 5.5 C **(d)** 1.2 C **(e)** 0.83 C

21–2 Resistance and Ohm's Law

Typically, current does not flow through a circuit unimpeded. The wires through which the current flows offer some **resistance** to this flow. The more resistance R there is in the wire, the smaller the current I that will flow for a given potential difference V across the wire. The relationship between these three quantities is referred to as **Ohm's law**:

$$V = IR$$

The resistance R has the SI unit V/A; this unit is called an **ohm** (Ω): 1 Ω = 1 V/A.

The amount of resistance in a particular piece of wire depends on the size and shape of the wire as well as the type of material. Resistance increases in direct proportion to the length L of the wire and decreases in inverse proportion to the cross-sectional area A of the wire. The dependence on the type of material is contained within a measured quantity called the **resistivity,** ρ. Putting this information together gives

$$R = \rho\frac{L}{A}$$

The resistivity of a material has the SI unit Ω·m.

Example 21–1 Flashlight Assume that the filament in the light bulb of a flashlight is made of a small piece of metal alloy. If this piece of metal has a radius of 0.500 mm, a length of 1.00 cm, and resistivity of 4.39×10^{-4} Ω·m, what current runs through this filament if the flashlight operates on two 1.50-volt batteries?

Picture the Problem The diagram shows the circuit for the flashlight.

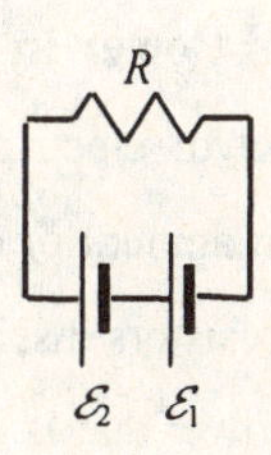

Strategy The current can be determined from Ohm's law once the resistance is known. The resistance can be calculated from the given information.

Solution

1. The resistance can be determined from the resistivity: $R = \rho\frac{L}{A} = \rho\frac{L}{\pi r^2}$

2. The net potential difference across the resistor: $\mathcal{E} = \mathcal{E}_1 + \mathcal{E}_2$

3. By Ohm's law we have: $I = \frac{\mathcal{E}}{R} = \frac{\pi r^2(\mathcal{E}_1 + \mathcal{E}_2)}{\rho L}$

4. The numerical result is: $I = \frac{\pi(0.500 \times 10^{-3}\ \text{m})^2(3.00\ \text{V})}{(4.39 \times 10^{-4}\ \Omega \cdot \text{m})(1.00 \times 10^{-2}\ \text{m})} = 0.537\ \text{A}$

Insight Notice that the two batteries oriented with the same polarity provide an emf equal to the sum of the individual emfs.

Practice Quiz

2. A potential difference V is applied across a resistor of resistance R producing a current I through the resistor. If a potential difference of $2V$ is applied across the same resistor, what current will flow through it?

   (a) $I/4$ **(b)** $I/2$ **(c)** I **(d)** $2I$ **(e)** $4I$

3. A potential difference is applied across a resistor of resistance R, producing a current I through the resistor. If the resistor is replaced by one of resistance $2R$, what current will flow through it?

(a) $I/4$ **(b)** $I/2$ **(c)** I **(d)** $2I$ **(e)** $4I$

4. A potential difference V is applied across a resistor, made from a copper wire of length L and radius r, producing a current I through the resistor. If the resistor is replaced by another copper wire of equal length and radius $2r$, what current will flow through it?

(a) $I/4$ **(b)** $I/2$ **(c)** I **(d)** $2I$ **(e)** $4I$

21–3 Energy and Power in Electric Circuits

From the definition of electric potential (potential energy per unit charge), we know that energy is transferred when an amount of charge moves through a potential difference. Batteries transfer this energy to the charge, and resistors dissipate this energy in the form of heat. The rate at which this energy transfer takes place is the power that is either generated or dissipated by a device in the circuit. In general, when a current I flows as a result of a potential difference V, the electrical power used is given by

$$P = IV$$

Recall that the SI unit of power is the watt: 1 W = 1 A·V = 1 J/s. When this power is being dissipated through a resistance R, we can apply Ohm's law ($V = IR$) to find other useful relationships for the power. Specifically, it is straightforward to show that

$$P = IV = I^2R = V^2/R$$

Because power is energy divided by time, then power times time will give energy. A unit of energy that is commonly used to track the household use of electricity is the kilowatt-hour, kWh. This quantity equals the amount of energy used if 1 kW of power is consumed for one hour. In joules

$$1\ \text{kWh} = 3.6 \times 10^6\ \text{J}$$

Example 21–2 Power (a) Determine the power requirements of the flashlight in Example 21-1. **(b)** If the flashlight is left on for a half hour, how much energy in kilowatt-hours is used?

Picture the Problem The diagram shows the circuit for the bulb in the flashlight.

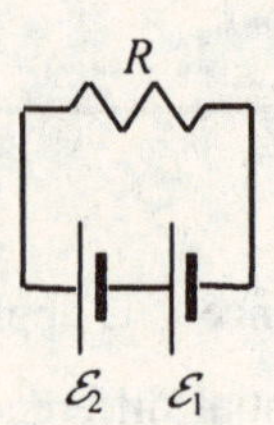

Strategy For part (a) we can use the result of Example 21-1 to obtain the power. For part (b) the energy can be found as the power times the time.

Solution

Part (a)

1. Use the current from Example 21–1 to calculate the power: $P = IV = (0.5367\ \text{A})(3.00\ \text{V}) = 1.61\ \text{W}$

Part (b)

2. The energy in kilowatt-hours is: $E = Pt = (0.001610\ \text{kW})[0.500\ \text{h}] = 8.05\times10^{-4}\ \text{kWh}$

Insight In part (b), we converted the power to kilowatts and used the time in hours to directly get the result in the desired units.

Practice Quiz

5. What is the resistance of a device that consumes 100 W of power when connected to a 120-V source?

   (a) 144 Ω **(b)** 12,000 Ω **(c)** 1.2 Ω **(d)** 0.83 Ω **(e)** 20 Ω

6. How much energy is consumed if a 75-W light bulb is left on for 12 hours?

   (a) 900 kWh **(b)** 6.3 kWh **(c)** 0.90 kWh **(d)** 160 kWh **(e)** 0.16 kWh

21–4 Resistors in Series and Parallel

A resistor in a circuit, indicated by a jagged line, represents a circuit element (such as a light bulb or a heater) that contains resistance. Circuit elements can be connected in different ways, and the overall resistance, or **equivalent resistance**, of the combination depends on how they are connected. In this section, we learn how to determine the equivalent resistance for combinations of resistors connected in **series** and in **parallel.**

Resistors in series are connected one after the other (or end to end). Connecting resistors in series has the same effect as making one resistor longer. It is important to remember that

the same current flows through resistors that are in series with each other.

The result is that the equivalent resistance, R_{eq}, of N resistors connected in series is the sum of the individual resistances

$$R_{eq} = R_1 + R_2 + \cdots + R_N = \sum_{i=1}^{N} R_i$$

Consequently, R_{eq} is greater than any of the individual resistances that contribute to it.

Resistors in parallel are connected across the same potential difference.

Connecting resistors in parallel has the same effect as making one resistor wider. The result is that the equivalent resistance of N resistors, connected in parallel, can be found by the following relation:

$$\frac{1}{R_{eq}} = \frac{1}{R_1} + \frac{1}{R_2} + \cdots + \frac{1}{R_N} = \sum_{i=1}^{N} \frac{1}{R_i}$$

Once $1/R_{eq}$ is determined, we can take the reciprocal of this result to get the value of R_{eq} in ohms. As a consequence, we can see that for resistors in parallel, R_{eq} is smaller than the smallest of the individual resistances that contribute to it.

Example 21–3 Equivalent Resistance Determine the equivalent resistance of the circuit shown, given that $R_1 = 3.3\ \Omega$, $R_2 = 4.5\ \Omega$, $R_3 = 6.1\ \Omega$, and $R_4 = 2.8\ \Omega$.

Picture the Problem The diagram shows the circuit in question.

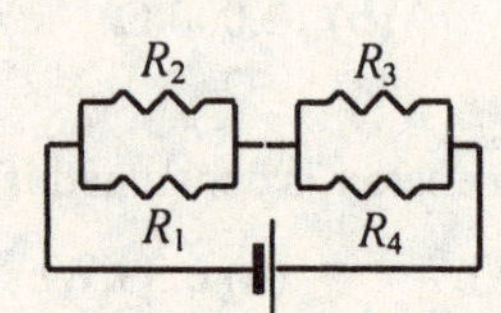

Strategy We will first get the equivalent resistance of each parallel combination, and then combine those results in series.

Solution

1. The equivalent resistance of R_1 and R_2 is: $R_{12} = \left[\frac{1}{R_1} + \frac{1}{R_2}\right]^{-1} = \left[\frac{1}{3.3\ \Omega} + \frac{1}{4.5\ \Omega}\right]^{-1} = 1.904\ \Omega$

2. The equivalent resistance of R_3 and R_4 is: $R_{34} = \left[\frac{1}{R_3} + \frac{1}{R_4}\right]^{-1} = \left[\frac{1}{6.1\ \Omega} + \frac{1}{2.8\ \Omega}\right]^{-1} = 1.919\ \Omega$

3. The overall equivalent resistance then is: $R_{eq} = R_{12} + R_{34} = 1.904\ \Omega + 1.919\ \Omega = 3.8\ \Omega$

Insight With many resistors, it is often useful to take an "inside-out" approach as shown here.

Practice Quiz

7. A 2.0-Ω resistor is connected in series with a 3.0-Ω resistor. If this series combination is then connected in parallel with a third resistor of resistance 2.5 Ω, what is the equivalent resistance of this three-resistor circuit?

   (a) 2.5 Ω **(b)** 5.0 Ω **(c)** 1.7 Ω **(d)** 3.9 Ω **(e)** 7.5 Ω

8. A resistor, $R_1 = 3.0\ \Omega$, is connected in series with a resistor, $R_2 = 5.0\ \Omega$. When this combination is connected to a battery, which resistor consumes more power?

(a) R_1 **(b)** R_2 **(c)** They consume the same amount of power.

9. A resistor, $R_1 = 3.0\ \Omega$, is connected in parallel with a resistor, $R_2 = 5.0\ \Omega$. When this combination is connected to a battery, which resistor consumes more power?

(a) R_1 **(b)** R_2 **(c)** They consume the same amount of power.

21–5 Kirchhoff's Rules

As circuitry becomes more complicated, it becomes difficult to analyze the circuit using simple applications of Ohm's law. For these we use **Kirchhoff's rules**, which apply the conservation of electric charge and the conservation of energy to circuit analysis.

Kirchhoff's first rule (for the conservation of charge) is called the **junction rule**:

The algebraic sum of all currents meeting at any junction in a circuit must equal zero.

A **junction** is a point in a circuit where three or more wires meet so that the current in the circuit may take different paths into or out of this point. In applying the junction rule as stated above, currents *entering* the junction are given a positive sign, and currents *leaving* the junction are given a negative sign. Application of the junction rule seems to imply that the directions of the currents must be known; however, it is not always obvious what the direction of every current will be. Fortunately, in these cases it is sufficient just to guess a direction and solve the equations. If the value of the current whose direction you guessed works out to be negative, then you guessed wrong and you therefore know that it actually flows in the opposite direction from what you guessed.

Kirchhoff's second rule (for the conservation of energy) is called the **loop rule**:

The algebraic sum of all potential differences around any closed loop in a circuit must equal zero.

Applying these two rules involves making several choices. These choices include the following:

1. choose the directions of the currents in different parts of the circuit
2. choose which junctions and loops you will use for your analysis
3. choose directions for traversing the loops to generate your loop-rule equations

For step 3, there are several things you should remember when writing your loop-rule equations.

- Crossing a battery from the negative terminal to the positive terminal is a potential increase, and crossing it from positive to negative is a potential decrease.
- Crossing a resistor in the same direction as the chosen current is a potential decrease.
- Crossing a resistor in the direction opposite the chosen current is a potential increase.

Applying these rules carefully will give you a system of equations that can be solved for currents, resistances, and potential differences in circuits.

Exercise 21–4 Kirchhoff's Rules Use Kirchhoff's rules to determine the overall current and the current through each of the resistors in the circuit shown. The relevant data is that $R_1 = 2.5\ \Omega$, $R_2 = 3.3\ \Omega$, $R_3 = 1.8\ \Omega$, $\mathcal{E}_1 = 20$ V, and $\mathcal{E}_2 = 6.0$ V.

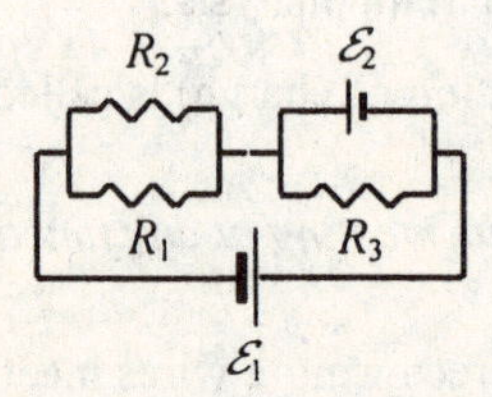

Solution

First, we choose the directions for the currents in the circuit realizing that some or all of these choices may not be correct. The figure below also labels the junctions in the circuit **A** to **D** (note that **B** and **C** are the same junction).

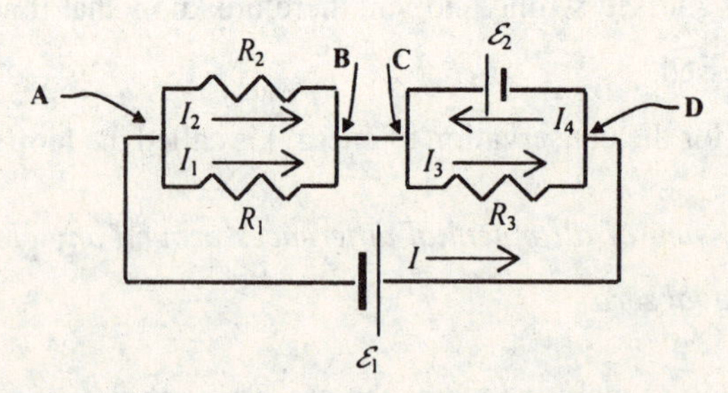

By examining the circuit, we see that I_3 can be determined by applying the loop rule to the small loop involving R_3 and $\mathcal{E}_2$ shown below.

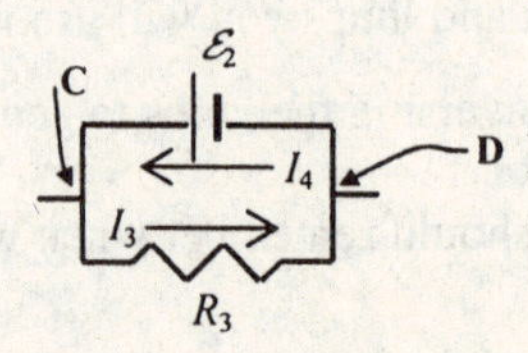

Moving around this loop counterclockwise from junction **D** back to **D** produces the loop-rule equation

$$\mathcal{E}_2 - I_3 R_3 = 0$$

which gives

$$I_3 = \frac{\mathcal{E}_2}{R_3} = \frac{6.0\ \text{V}}{1.8\ \Omega} = 3.3\ \text{A}$$

The fact that we got a positive value for I_3 shows that we chose the correct direction.

Further examination suggests that I_2 can be determined by applying the loop rule to the large loop that includes $\mathcal{E}_1$ and $\mathcal{E}_2$ and R_2.

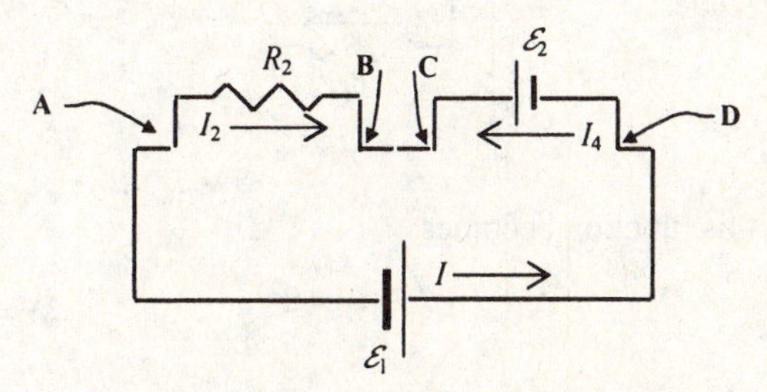

Moving completely around this loop counterclockwise produces the loop-rule equation

$$\mathcal{E}_1 + \mathcal{E}_2 + I_2 R_2 = 0$$

which gives

$$I_2 = -\frac{\mathcal{E}_1 + \mathcal{E}_2}{R_2} = -\frac{20\ \text{V} + 6.0\ \text{V}}{3.3\ \Omega} = -7.88\ \text{A}$$

The negative sign means that we chose the wrong direction for I_2.

Let's pause for a moment to notice that I_1 can be found using three different loops: (1) One loop would be the parallel combination involving R_1 and R_2 only, (2) another loop would be the larger path including $\mathcal{E}_1$, $\mathcal{E}_2$, and R_1, and (3) the third loop would come from the path including $\mathcal{E}_1$, R_3, and R_1. The choice with the least chance for error is (2). Why?

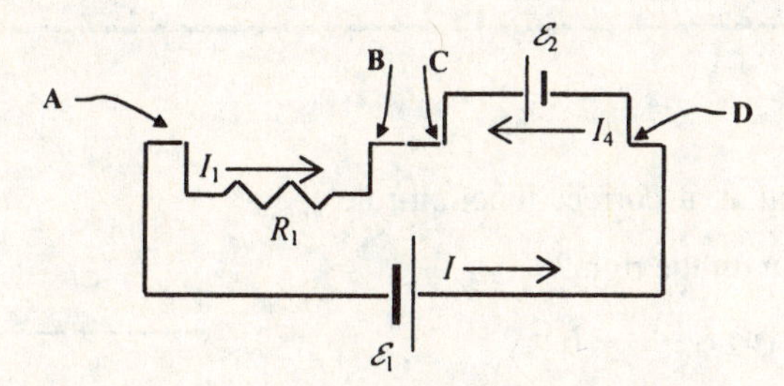

Moving around this loop counterclockwise we get the loop-rule equation

$$\mathcal{E}_1 + \mathcal{E}_2 + I_1 R_1 = 0$$

which gives

$$I_1 = -\frac{\mathcal{E}_1 + \mathcal{E}_2}{R_1} = -\frac{20\text{ V} + 6.0\text{ V}}{2.5\ \Omega} = -10.4\text{ A}$$

You can see that the wrong direction was also chosen for I_1.

All that remains is to find the overall current I. Inspection of the circuit suggests that we can determine I by applying the junction rule at junction **A**.

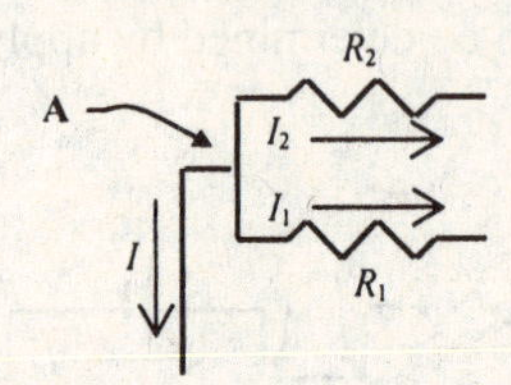

The junction-rule equation at this junction becomes

$$-I - I_1 - I_2 = 0$$

This gives

$$I = -(I_1 + I_2) = -(-10.4\text{ A} - 7.88\text{ A}) = 18\text{ A}$$

Thus, our final results are $I_1 = 10\text{ A},\quad I_2 = 7.9\text{ A},\quad I_3 = 3.3\text{ A}$, and $I = 18\text{ A}$

Make sure you understand the sign associated with each term in the loop-rule and junction-rule equations; it is very important that these be done consistently and correctly. Also, keep in mind that when you get a negative result for a current, that negative result must be used in any subsequent mathematical step; only drop the negative when stating the final result for that current. It is always better if you can choose the correct current directions in the beginning. In the circuit for this problem, the polarities of the two emfs would immediately suggest that the directions of I_1 and I_2 should be opposite to what was chosen. Try to see this fact. The choice made for this example was done only to show you what happens when the wrong choice is made.

Practice Quiz

10. Which of the following is a correct junction-rule equation for the diagram on the right?

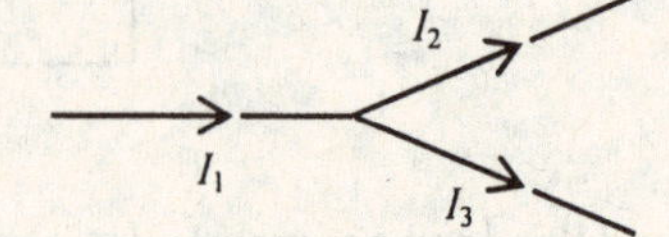

(a) $I_1 + I_2 + I_3 = 0$

(b) $I_1 + I_2 - I_3 = 0$

(c) $-I_1 - I_2 + I_3 = 0$

(d) $I_1 - I_2 - I_3 = 0$

(e) $I_1 - I_2 + I_3 = 0$

11. Which of the following is a correct loop-rule equation for the diagram on the right?

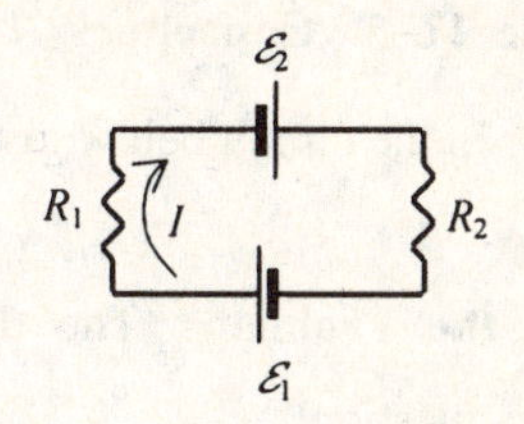

(a) $\mathcal{E}_1 + IR_1 + \mathcal{E}_2 + IR_2 = 0$

(b) $\mathcal{E}_1 + IR_1 - \mathcal{E}_2 + IR_2 = 0$

(c) $\mathcal{E}_1 - IR_1 + \mathcal{E}_2 + IR_2 = 0$

(d) $-\mathcal{E}_1 - IR_1 + \mathcal{E}_2 + IR_2 = 0$

(e) $\mathcal{E}_1 - IR_1 + \mathcal{E}_2 - IR_2 = 0$

21–6 Circuits Containing Capacitors

Capacitors are commonly used in circuits and can be combined in series and parallel to produce an overall equivalent capacitance C_{eq} that depends on the type of connection. As with resistors,

capacitors in parallel are connected across the same potential difference.

In the case of parallel-plate capacitors ($C = \varepsilon_0 A/d$), connecting capacitors in parallel is equivalent to making the plates of one capacitor larger. The result is that for N capacitors connected in parallel the equivalent capacitance is

$$C_{eq} = C_1 + C_2 + \cdots + C_N = \sum_{i=1}^{N} C_i$$

Therefore, in parallel, the equivalent capacitance is larger than any of the individual capacitances that contribute to it.

Just as resistors in series have the same current flowing through them, capacitors in series have the same current flow onto them, and thus each capacitor stores the same amount of charge Q. Therefore,

capacitors in series store equal amounts of charge.

Connecting capacitors in series has the effect of increasing the distance between the plates of one capacitor. The result for N capacitors connected in series is that the equivalent capacitance can be determined by

$$\frac{1}{C_{eq}} = \frac{1}{C_1} + \frac{1}{C_2} + \cdots + \frac{1}{C_N} = \sum_{i=1}^{N} \frac{1}{C_i}$$

Once $1/C_{eq}$ is determined, we can take the reciprocal to determine C_{eq} in farads. Therefore, in series, the equivalent capacitance is smaller than the smallest of the individual capacitances that contribute to it.

Example 21–5 Capacitors Determine both the charge on and the potential difference across each capacitor in the circuit below given that $C_1 = 12\ \mu\text{F}$, $C_2 = 5.0\ \mu\text{F}$, $C_3 = 6.5\ \mu\text{F}$, and $\mathcal{E} = 12$ V.

Picture the Problem The diagram shows the circuit in question.

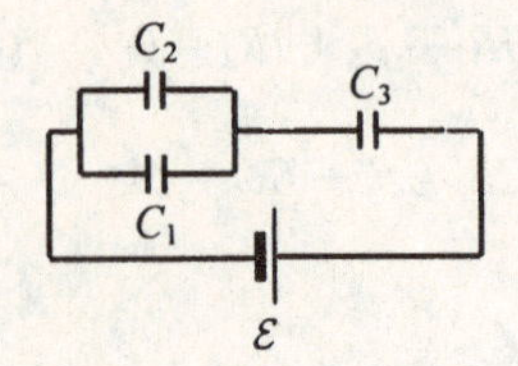

Strategy We can use the known facts about capacitors in series and parallel to obtain the required information.

Solution

1. The equivalent capacitance of C_1 and C_2 is: $C_{12} = C_1 + C_2 = 12\mu\text{F} + 5.0\mu\text{F} = 17.0\mu\text{F}$

2. The overall equivalent capacitance is: $C_{eq} = \left[\frac{1}{C_{12}} + \frac{1}{C_3}\right]^{-1} = \left[\frac{1}{17.0\mu\text{F}} + \frac{1}{6.5\mu\text{F}}\right]^{-1} = 4.702\ \mu\text{F}$

3. The total charge stored is: $Q_{tot} = C_{eq}V = (4.702\ \mu\text{F})(12\ \text{V}) = 56.43\ \mu\text{C}$

4. Since the same charge is stored on capacitors in series, the charge stored on C_3 is: $Q_3 = Q_{tot} = 56\ \mu\text{C}$

5. The potential difference across C_3 is: $V_3 = \frac{Q_3}{C_3} = \frac{56.43\ \mu\text{C}}{6.5\ \mu\text{C}} = 8.7\ \text{V}$

6. By the loop rule, the potential difference across the C_1 and C_2 combination is: $V_{12} = \mathcal{E} - V_3 = 12\ \text{V} - 8.681\ \text{V} = 3.319\ \text{V}$

7. Since capacitors in parallel are connected across the same potential difference: $V_1 = V_2 = 3.3\ \text{V}$

8. The charge stored on C_1 is: $Q_1 = C_1V_1 = (12\ \mu\text{F})(3.319\ \text{V}) = 40\ \mu\text{C}$

9. The charge stored on C_2 is: $Q_2 = C_2V_2 = (5.0\ \mu\text{F})(3.319\ \text{V}) = 17\ \mu\text{C}$

Insight Make sure you understand the results of steps 4 and 7; these steps are the keys to the solution.

Practice Quiz

12. A 6.2-μF capacitor is connected in parallel with a 2.4-μF capacitor. If this parallel combination is then connected in series with a third capacitor of capacitance 3.8 μF, what is the equivalent capacitance of this three-capacitor circuit?

(a) 5.5 μF **(b)** 1.7 μF **(c)** 8.6 μF **(d)** 12.4 μF **(e)** 2.6 μF

13. Given two capacitors of equal capacitance, which type of connection would allow them to store the most combined charge?

(a) series **(b)** parallel **(c)** Both connections would store the same amount of charge.

21–7 RC Circuits

For our purposes, an **RC circuit** is one that contains resistance and capacitance in series. In these circuits, the charge on the capacitor builds up (and dies off) slowly in comparison to what happens in circuits containing only resistance and no capacitance. For an RC circuit in which the capacitor is *initially uncharged*, the charge builds up on the capacitor according to the equation

$$q(t) = C\mathcal{E}\left(1 - e^{-t/\tau}\right) \qquad \text{[charging]}$$

where $\mathcal{E}$ is the source emf. The quantity τ is given by $\tau = RC$ and is called the **time constant** of the circuit. The time constant provides a characteristic amount of time for the charge to build because after an amount of time $t = \tau$, the capacitor builds up (63.2%) to its maximum charge of $q_{\max} = C\mathcal{E}$. In the present case of an initially uncharged capacitor, the current starts out at its maximum value and falls off exponentially:

$$I(t) = \frac{\mathcal{E}}{R} e^{-t/\tau} \qquad \text{[charging]}$$

If the capacitor *is initially charged* with a charge Q, has a potential difference V across it, and is allowed to discharge, both the charge and current fall off exponentially:

$$q(t) = Qe^{-t/\tau}; \quad I(t) = \frac{V}{R} e^{-t/\tau} \qquad \text{[discharging]}$$

Example 21–6 An RC Circuit Determine the time constant of the RC circuit shown below.

Picture the Problem The diagram shows the circuit in question.

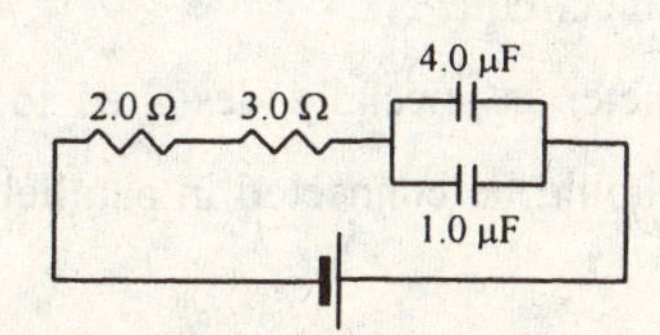

Strategy We need to determine the correct values of R_{eq} and C_{eq} and then use them to get τ.

Solution

1. The equivalent resistance is: $R_{eq} = R_1 + R_2 = 2.0\ \Omega + 3.0\ \Omega = 5.0\ \Omega$

2. The equivalent capacitance is: $C_{eq} = C_1 + C_2 = 4.0\ \mu\text{F} + 1.0\ \mu\text{F} = 5.0\ \mu\text{F}$

3. The time constant is: $\tau = R_{eq}C_{eq} = (5.0\ \Omega)(5.0\ \mu\text{F}) = 25\ \mu\text{s}$

Insight Here, despite the fact that there are two resistors and two capacitors, the resistance and capacitance are still in series.

Practice Quiz

14. Which of the following statements is true for charging a capacitor?

(a) The time constant is how long it takes the charge to build up to 36.8% of its maximum value.

(b) The time constant is how long it takes the current to build up to 63.2% of its maximum value.

(c) The time constant is how long it takes the current to build up to 36.8% of its maximum value.

(d) The time constant is how long it takes the charge to build up to 63.2% of its maximum value.

(e) The time constant is how long it takes the current to fall off to 63.2% of its maximum value.

15. Which of the following statements is true for discharging a capacitor?

(a) The time constant is how long it takes the charge to fall off to 36.8% of its maximum value.

(b) The time constant is how long it takes the current to build up to 63.2% of its maximum value.

(c) The time constant is how long it takes the current to build up to 36.8% of its maximum value.

(d) The time constant is how long it takes the charge to fall off to 63.2% of its maximum value.

(e) The time constant is how long it takes the current to fall off to 63.2% of its maximum value.

*21–8 Ammeters and Voltmeters

An **ammeter** is a device specially made for measuring currents in a circuit. An ammeter should be connected in series with the device whose current is sought. Ideally, an ammeter should have zero resistance. In practice, the resistance of an ammeter should be much less than the resistances of other devices in the circuit.

A voltmeter is specially designed to measure potential differences across devices in a circuit. A voltmeter should be connected in parallel with the device across which the potential difference is being sought. Ideally, a voltmeter should have infinite resistance. In practice, the resistance of a voltmeter should be much greater than the resistances of other devices in the circuit.

Reference Tools and Resources

I. Key Terms and Phrases

electric current results from the flow of electric charge

electric circuits closed paths containing circuit elements through which current can flow

direct current (DC) circuits circuits in which the current always flows in one direction

electric battery a device, that maintains a potential difference and is used as an energy source for electric circuits

electromotive force (emf) the potential difference across an ideal battery

resistance the opposition to the flow of charge through a wire due to the properties of the wire

ohm the SI unit of resistance

resistivity the property of a substance that partially determines the resistance of objects made from it

connected in series circuit elements connected one after the other such that the same current flows through them

connected in parallel circuit elements connected across the same potential difference

Kirchhoff's rules two rules that apply the conservation of charge (junction rule) and the conservation of energy (loop rule) to electric circuits

junction a point in a circuit where three or more wires meet so that the current in the circuit may take different paths into or out of this point

loop any closed path in a circuit

time constant the characteristic amount of time, $\tau = RC$, for an initially uncharged capacitor to charge up to 63.2% of its maximum value in an *RC* circuit

ammeter a device designed to measure the current through a circuit element

voltmeter a device designed to measure the potential difference across a circuit element

II. Important Equations

Name/Topic	Equation	Explanation
Ohm's law	$V = IR$	The relationship between potential difference, current, and resistance
resistivity	$R = \rho \frac{L}{A}$	How the resistance of a wire depends on the properties of the wire
electrical power	$P = IV = I^2R = V^2/R$	The electrical power transformed by a device in a circuit
resistors in series	$R_{eq} = R_1 + R_2 + \cdots + R_N$	How resistors in series combine
resistors in parallel	$\frac{1}{R_{eq}} = \frac{1}{R_1} + \frac{1}{R_2} + \cdots + \frac{1}{R_N}$	How resistors in parallel combine
capacitors in parallel	$C_{eq} = C_1 + C_2 + \cdots + C_N$	How capacitors in parallel combine
capacitors in series	$\frac{1}{C_{eq}} = \frac{1}{C_1} + \frac{1}{C_2} + \cdots + \frac{1}{C_N}$	How capacitors in series combine

III. Know Your Units

Quantity	Dimension	SI Unit
electric current (I)	[A]	A
resistance (R)	$[M][L^2][A^{-2}][T^{-3}]$	Ω
resistivity (ρ)	$[M][L^3][A^{-2}][T^{-3}]$	Ω·m

Puzzle

BULB MARKET

Houses, boats, and cars have electrical systems based on 120 V, 24 V, and 12 V respectively. Let's say we have a 60-watt bulb from each, and enough adapters to plug any bulb into any system. Which combination of bulb and system will give the most light? Which will give the least? Answer these questions in words, not equations, briefly explaining how you obtained your answers.

Answers to Selected Conceptual Questions and Exercises

Conceptual Questions

6. Car headlights are wired in parallel, as we can tell by the fact that some cars have only one working headlight.

12. Resistivity is an intrinsic property of a particular substance. In this sense it is similar to density, which has a particular value for each particular substance. Resistance, however, is a property associated with a given resistor. For example, the resistance of a given wire can be large because its resistivity is large, or because it is long. Similarly, the weight of a ball can be large because its density is large, or because it has a large radius.

18. The light shines brightest immediately after the switch is closed. With time, the intensity of the light diminishes. Eventually, the light stops glowing altogether.

Conceptual Exercises

10. Assuming the lights are connected to the same potential difference, V, the resistance of the lights can be compared using $P = V^2/R.$ **(a)** Since light A has the greater power rating, it follows that the resistance of light A is less than the resistance of light B. **(b)** Note that the power is inversely proportional to the resistance. It follows that the ratio of the resistance of light A to the resistance of light B is 1/4.

14. The fuse should be connected in series with the circuit it protects. Then, if the fuse burns out, no current will flow in the circuit.

28. When the upper and lower capacitors are connected in each circuit, the charge will redistribute to produce a voltage of the same magnitude across each capacitor. **(a)** The ranking in order of increasing charge on the left plate of the upper capacitor is B < A < C. **(b)** The ranking in order of increasing charge on the left plate of the lower capacitor is B < A = C.

Solutions to Selected End-of-Chapter Problems

15. Picture the Problem: A wire conducts current when a potential difference is applied between the two ends.

Strategy: Ohm's Law (Equation 21-2) gives the relationship between potential difference, current, and resistance for any circuit element. To solve this problem we can combine Equation 21-3 with Ohm's law to find the resistivity of the material from which the wire was made.

Solution: Combine Equations 21-2 and 21-3 to find ρ:

$$V = IR = I\rho\frac{L}{A} = I\rho\frac{L}{\frac{1}{4}\pi D^2}$$

$$\therefore \rho = \frac{\pi D^2 V}{4IL} = \frac{\pi\left(0.33\times10^{-3}\text{ m}\right)^2(12\text{ V})}{4(2.1\text{ A})(6.9\text{ m})} = \boxed{7.1\times10^{-8}\ \Omega\cdot\text{m}}$$

Insight: The resistivity of this material is a little higher than tungsten but lower than iron. If it is a pure metal we might suspect it is made out of one of the transition metals.

25. Picture the Problem: Two different rating schemes measure the energy stored in a car battery.

Strategy: The two rating schemes each measure the amount of current the battery can supply at a specified voltage for a period of time. Find the power produced by the battery by multiplying the current by the voltage (Equation 21-4) and then the energy delivered by multiplying the power by the time elapsed.

Solution: **1.** Find the energy delivered under the first rating scheme:

$$E_1 = P\Delta t = IV\Delta t = (905\text{ A})(7.2\text{ V})(30.0\text{ s}) = 2.0\times10^5\text{ J}$$

2. Repeat for the second rating scheme:

$$E_1 = P\Delta t = IV\Delta t = (25\text{ A})(10.5\text{ V})(155\text{ min}\times 60\text{ s/min}) = 2.4\times10^6\text{ J}$$

3. The $\boxed{\text{155-minute reserve capacity}}$ rating represents the greater amount of energy delivered by the battery.

Insight: During the cold cranking test the battery pours out a huge amount of current, but for a short period of time. It requires 12 times more energy to sustain the much lower current for over 2.5 hours!

37. Picture the Problem: Six resistors are connected in the manner indicated by the diagram at right.

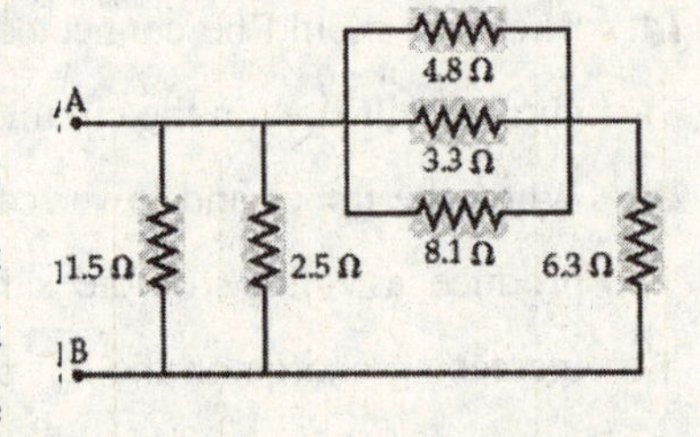

Strategy: Use the rules concerning resistors in series and in parallel to write an expression for the equivalent resistance of the entire circuit Begin by finding the equivalent resistance of the three uppermost resistors (4.8 Ω, 3.3 Ω, and 8.1 Ω), then add their equivalent resistance to the 6.3 Ω resistor. The equivalent resistance of those four can then be added to the 1.5Ω and 2.5 Ω resistors according to Equation 21-10 because the three are connected in parallel.

Solution: Add the uppermost 3 resistors in parallel, then add them in series with the 6.3 Ω resistor, then add the result in parallel with the 1.5 Ω and 2.5 Ω resistors to find R_{eq}.

$$\frac{1}{R_{eq}} = \frac{1}{1.5\Omega} + \frac{1}{2.5\Omega} + \frac{1}{6.3\Omega + \left(\frac{1}{4.8\Omega} + \frac{1}{3.3\Omega} + \frac{1}{8.1\Omega}\right)^{-1}}$$

$$R_{eq} = \boxed{0.84\ \Omega}$$

Insight: The equivalent resistance of the entire circuit is less than the smallest resistor (1.5 Ω) in the network.

51. Picture the Problem: Four resistors and two batteries are connected as shown in the circuit diagram at right.

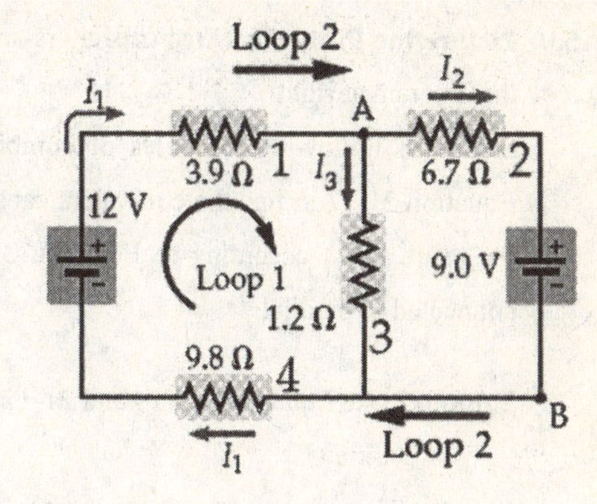

Strategy: The circuit can be analyzed by applying Kirchoff's rules. First apply the Junction Rule to point A in the circuit, then apply the Loop Rule to two loops, the left-hand loop 1 and the outside loop 2 labeled in the diagram. These three equations can be combined to algebraically find I_1, I_2, and I_3. From the currents we can find the potential difference between the points A and B.

Solution: 1. (a) Apply the Junction Rule to point A:

$$I_1 = I_2 + I_3 \ldots\ldots \text{(i)}$$

2. Apply the Loop Rule to loop 1, beginning in lower left-hand corner, and solve for I_3:

$$0 = 12\text{ V} - I_1R_1 - I_3R_3 - I_1R_4$$

$$I_3 = \frac{12\text{ V}}{R_3} - \frac{R_1 + R_4}{R_3}I_1 = 10\text{ A} - \frac{13.7\ \Omega}{1.2\ \Omega}I_1 \ldots\ldots \text{(ii)}$$

3. Apply the Loop Rule to loop 2, beginning in lower left-hand corner, and solve for I_2:

$$0 = 12\text{ V} - I_1R_1 - I_2R_2 - 9.0\text{ V} - I_1R_4$$

$$I_2 = \frac{12.0 - 9.0\text{ V}}{R_2} - \frac{R_1 + R_4}{R_2}I_1 = \left(\frac{3.0\text{ V}}{6.7\ \Omega}\right) - \left(\frac{13.7\ \Omega}{6.7\ \Omega}\right)I_1 \ldots\ldots \text{(iii)}$$

4. Substitute equations (ii) and (iii) into Equation (i).

$$I_1 = 10\text{ A} - \left(\frac{13.7}{1.2}\right)I_1 + \left(\frac{3.0}{6.7}\text{ A}\right) - \left(\frac{13.7}{6.7}\right)I_1$$

$$\therefore\ \left(1 + \frac{13.7}{1.2} + \frac{13.7}{6.7}\right)I_1 = 10\text{ A} + \frac{3}{6.7}\text{ A}$$

$$\therefore\ I_1 = \frac{10\text{ A} + \frac{3}{6.7}\text{ A}}{1 + \frac{13.7}{1.2} + \frac{13.7}{6.7}} = \underline{\underline{0.72\text{ A}}}$$

5. Substitute I_1 into equations (ii) and (iii):

$$I_3 = 10\text{ A} - \frac{13.7}{1.2}(0.72\text{ A}) = \underline{\underline{1.8\text{ A}}}$$

$$I_2 = \left(\frac{3.0}{6.7}\text{ A}\right) - \frac{13.7}{6.7}(0.72\text{ A}) = \underline{\underline{-1.0\text{ A}}}$$

6. The currents through each resistor are as follows: 3.9 Ω, 9.8 Ω: $\boxed{0.72\text{ A}}$; 1.2 Ω: $\boxed{1.8\text{ A}}$; 6.7 Ω: $\boxed{1.0\text{ A}}$.

7. The potential at point A is $\boxed{\text{greater than}}$ that at point B because I_3 flows in the direction shown in the diagram and produces a potential drop across R_3.

8. Find the potential drop across R_3:

$$V_A - V_B = I_3R_3 = (1.8\text{ A})(1.2\ \Omega) = \boxed{2.2\text{ V}}$$

Insight: Because we obtained a negative value for I_2, it must be flowing in the direction opposite that indicated by the arrow in the circuit diagram.

53. Picture the Problem: Three capacitors are connected together as shown in the diagram at right.

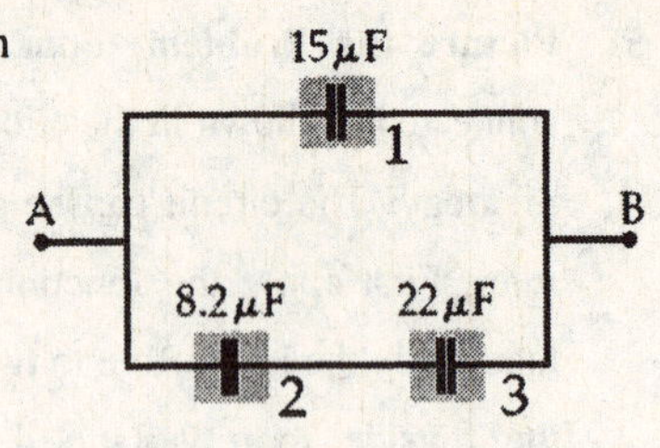

Strategy: Following the rules of combining capacitors, we can use Equation 21-17 to find the equivalent capacitance of C_2 and C_3, and add the result to C_1 according to Equation 21-14, because C_1 and C_{23} are connected in parallel.

Solution: Use Equations 21-14 and 21-17 to find C_{eq}:

$$C_{eq} = C_1 + \left(\frac{1}{C_2} + \frac{1}{C_3}\right)^{-1} = 15\ \mu\text{F} + \left(\frac{1}{8.2\ \mu\text{F}} + \frac{1}{22\ \mu\text{F}}\right)^{-1} = \boxed{21\ \mu\text{F}}$$

Insight: The equivalent capacitance of C_2 and C_3 is only about 6.0 μF, much smaller than the algebraic sum $8.2 + 22\ \mu\text{F} = 30\ \mu\text{F}$, because they are connected in series and combined according to Equation 21-17.

Answers to Practice Quiz

1. (b) **2.** (d) **3.** (b) **4.** (e) **5.** (a) **6.** (c) **7.** (c) **8.** (b) **9.** (a) **10.** (d) **11.** (e) **12.** (e) **13.** (b) **14.** (d) **15.** (a)

CHAPTER 22
MAGNETISM

Chapter Objectives

After studying this chapter, you should

1. know the rules for how magnetic poles interact and how to draw magnetic field lines.
2. be able to determine the magnetic force on a moving charge in a magnetic field.
3. be able to describe the motion of charged particles in uniform magnetic fields.
4. be able to calculate the force on a current-carrying wire, and the torque on a current loop, in a magnetic field.
5. be able to determine the magnetic field of a long, straight current, in the center of a current loop, and inside a solenoid.

Warm-Ups

1. The force on a charged particle in a magnetic field is very different from the force due to an electric field. Please list as many differences as you can. Don't forget to include differences in the direction as well as in the magnitude.
2. Estimate the force of the earth's magnetic field on a 10-cm segment of a typical wire in your home. (Hint: the magnitude of Earth's magnetic field is about 5.5×10^{-5} T.)
3. A proton moving downward enters a region of space with magnetic field $\vec{\mathbf{B}}$, which points eastward. In which direction is the force on the proton?

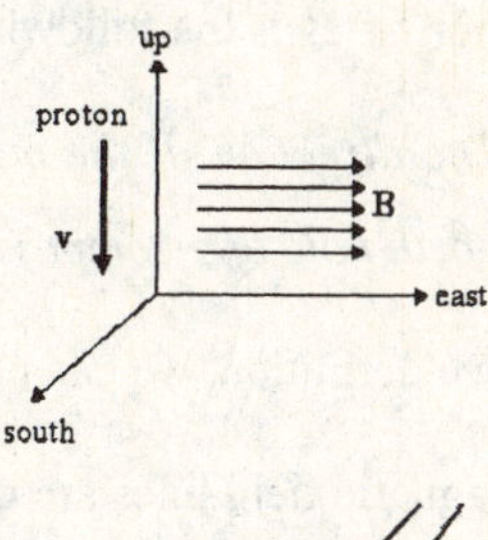

4. Two long parallel wires are separated by 0.2 meters and carry currents of 3 and 5 amps as shown in the figure. What is the direction of the magnetic force felt by the wire on the right?, By the wire on the left? How will the magnitudes of these forces compare?

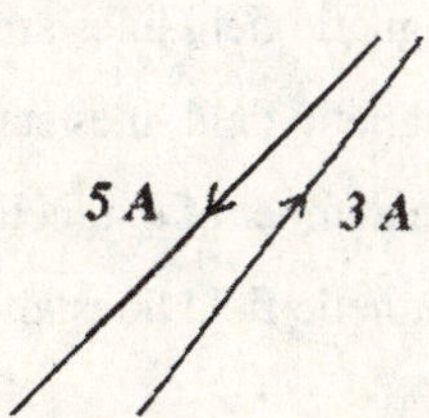

5. Let's say you shuffle across a carpet on a dry winter day and pick up a charge of 5 microcoulombs. What force will you feel due to the magnetic field of Earth?
6. You can make a good approximation to a "long solenoid" by buying a 50-yd spool of wire and winding it carefully (nice, even coils) around a cylindrical core (say a broomstick), then removing the core. Estimate the maximum magnetic field that can be produced in such a solenoid.

Chapter Review

In this chapter, we study magnetism. It turns out that electricity and magnetism are interconnected. We begin to glimpse at this connection in this chapter. The full story will emerge later.

22–1 The Magnetic Field

Some materials are almost always magnetic as a direct result of their structure; objects made of these materials are called **permanent magnets**. The prototype object for discussing the behavior of these materials is the simple bar magnet. The two ends of a bar magnet behave differently. One end, the **north pole**, of the magnet tends to point northward with respect to Earth. The other end, the **south pole**, of the magnet tends to point southward. Magnets always have both a north and a south pole. There is a force between two magnets. The basic behavior of this force is reminiscent of the force between two charges:

like magnetic poles repel; opposite poles attract.

As with electricity, magnetism is associated with the presence of a **magnetic field** for which we use the symbol $\vec{\mathbf{B}}$. Magnetic field lines can be drawn to get a visual representation of this field. In order to draw magnetic field lines, we must know how to determine the direction of the magnetic field. The rule for the direction of $\vec{\mathbf{B}}$ is the following:

The direction of the magnetic field at a given location is the direction that the north pole of a compass needle would point if placed at that location.

With the above definition, we can now state the rules for drawing magnetic field lines:

1. Magnetic field lines are tangent to the magnetic field at every point.
2. Magnetic field lines start on the north pole of a magnet and end on its south pole.
3. The number of magnetic field lines is proportional to the magnitude of the field.
4. Magnetic field lines always form closed loops.

See Figure 22–4 of the text for an example of magnetic field lines.

The fact that bar magnets interact with Earth is evidence that Earth has a magnetic field. The poles of Earth's magnetic field are near Earth's geographic poles. However, the definition of the magnetic poles and how they behave requires that the magnetic pole of the earth that's near Earth's north pole (called the **north magnetic pole**) is actually the *south pole* of Earth's magnetic field. Similarly, the magnetic pole of the earth that's near to Earth's south pole (called the **south magnetic pole**) is actually the *north pole* of Earth's magnetic field (see Figure 22–6 of the text).

Practice Quiz

1. If a magnetic field is directed along the negative y axis, toward which direction would the south pole of a compass needle point if placed in this magnetic field?
   (a) $+x$ axis **(b)** $-x$ axis **(c)** $+y$ axis **(d)** $-y$ axis **(e)** none of the above

22–2 The Magnetic Force on Moving Charges

In addition to the behavior described above, magnets also apply a force to charged objects that are moving through the magnetic field. The magnitude of the force on this charge depends on the magnitude of the charge q, the magnitude of its velocity v, the magnitude of the magnetic field B, and the smallest angle θ between the vectors $\vec{\mathbf{B}}$ and $\vec{\mathbf{v}}$:

$$F = qvB\sin\theta$$

Notice that if $\theta = 0°$ (or $180°$), that is, if the charge moves along the direction of the magnetic field, the force on it is zero. Therefore, a force is exerted only if the velocity of the charge has a component perpendicular to the magnetic field.

The above expression for the force on a moving charge can be used to obtain the magnitude (or strength) of the magnetic field:

$$B = \frac{F}{qv\sin\theta}$$

This relation shows that the SI unit of the magnetic field can be derived from N/(C·m/s). After some rearranging, this unit can be written as N/(A·m), which is called a **tesla** (T): 1 T = 1 N/A·m.

The direction of the force on a moving charge is found to be perpendicular to both $\vec{\mathbf{B}}$ and $\vec{\mathbf{v}}$. That is, $\vec{\mathbf{F}}$ is perpendicular to the plane formed by the vectors $\vec{\mathbf{B}}$ and $\vec{\mathbf{v}}$. However, this statement still leaves two possible directions for $\vec{\mathbf{F}}$ on either side of the $\vec{\mathbf{B}}$-$\vec{\mathbf{v}}$ plane. To determine the direction of $\vec{\mathbf{F}}$ more precisely we use the **magnetic force right-hand-rule**:

To find the direction of the magnetic force on a positive charge, point the fingers of your right hand in the direction of $\vec{\mathbf{v}}$. Orient your hand such that your fingers can curl toward the direction of $\vec{\mathbf{B}}$. Your thumb then indicates the direction of $\vec{\mathbf{F}}$. If the charge is negative, the force points in the opposite direction indicated by your thumb.

Essentially, the above rule determines on which side of the $\vec{\mathbf{B}}$-$\vec{\mathbf{v}}$ plane $\vec{\mathbf{F}}$ points; since you know $\vec{\mathbf{F}}$ must be perpendicular to this plane, then you know the precise direction of $\vec{\mathbf{F}}$.

The magnetic force on a moving charge is necessarily a three-dimensional situation. On two-dimensional sheets of paper this force involves the directions directly into the page and out of the page. These directions are indicated by using the symbol ⊗ for *into the page*, and ⊙ for *out of the page*.

Example 22–1 Force on a Moving Charge The magnetic field in a region is $\vec{\mathbf{B}} = (2.5\text{ T})\hat{\mathbf{x}} + (2.5\text{ T})\hat{\mathbf{y}}$. A particle of charge −1.7 C moves into this region with a velocity of $\vec{\mathbf{v}} = (4.5\text{ m/s})\hat{\mathbf{x}}$. Determine the magnitude and direction of the force on this charge.

Picture the Problem The sketch shows the magnetic field and velocity vectors.

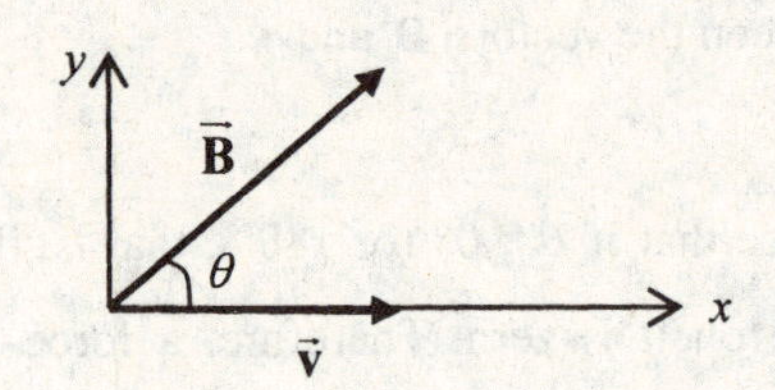

Strategy From the components given, it is clear that $\vec{\mathbf{B}}$ makes a 45° angle in the 1st quadrant. With this insight, we can proceed using the expression for the force.

Solution

1. The magnitude of $\vec{\mathbf{B}}$ is: $B = \sqrt{B_x^2 + B_y^2} = \sqrt{2(2.5\text{ T})^2} = 3.54\text{ T}$
2. The magnitude of the force is: $F = qvB\sin\theta = (1.7\text{ C})(4.5\tfrac{\text{m}}{\text{s}})(3.54\text{ T})\sin(45^\circ) = 19\text{ N}$
3. The right-hand rule gives the direction as: into the page, ⊗

Insight Only the magnitude of the charge was used to get the magnitude of the force as appropriate. Also, make sure you recognize that because the charge is negative, the direction of the force is opposite to the direction indicated by your thumb when you use the magnetic force right-hand rule.

Practice Quiz

2. A magnetic field is directed vertically from the bottom of this page toward the top, ↑. A positively charged particle moves with an initial velocity from the left toward the right, →. What is the direction of the force on this particle?

   (a) ↓ **(b)** ← **(c)** ⊗ **(d)** ⊙ **(e)** $F = 0$, so there is no direction.

3. An *electron* moves with a velocity that is into the page, ⊗. It experiences a magnetic force to the left, ←. What is the direction of the magnetic field in which the electron is moving?

   (a) ↓ **(b)** ↑ **(c)** ⊙ **(d)** → **(e)** $B = 0$, so there is no direction.

4. A particle of charge q moves with speed v at an angle θ with respect to a magnetic field $\vec{\mathbf{B}}$, producing a force of magnitude F on the particle. If a different particle of charge $-q$ moves with speed $v/2$ at the same angle θ through this same magnetic field, the magnitude of the force on this charge will be

   (a) $F/4$ **(b)** $F/2$ **(c)** F **(d)** $2F$ **(e)** $4F$

22–3 The Motion of Charged Particles in a Magnetic Field

For uniform magnetic fields, the force law for moving charges gives rise to three types of motion. One of these types is constant-velocity motion. This motion occurs when the velocity of the charge is either parallel or opposite to the direction of the magnetic field. For this situation, the magnetic force on the charge is zero, which can be seen from the fact that, for $\theta = 0°$ or $180°$, $\sin(\theta) = 0$.

Another type of motion that results in a uniform magnetic field is that of uniform circular motion. This motion occurs when the velocity of the charge is perpendicular to the magnetic field. Because the magnetic force is perpendicular to the velocity, so is the acceleration. When the acceleration of an object remains perpendicular to its velocity, uniform circular motion is the result. For a charged particle of mass m, whose charge has magnitude q, moving with velocity $\vec{\mathbf{v}}$ in a direction perpendicular to $\vec{\mathbf{B}}$, the radius of its uniform circular motion is given by

$$r = \frac{mv}{qB}$$

This radius is sometimes called the *cyclotron radius*.

The third type of motion is a combination of the above linear and circular motions. This occurs when the velocity of the charge is neither along nor perpendicular to the magnetic field lines. The component of the velocity that is perpendicular to $\vec{\mathbf{B}}$, $v_{\perp}$, contributes a circular component to its motion

$$r = \frac{mv_{\perp}}{qB}$$

and the fact that the velocity has a component along the magnetic field lines, $v_{\parallel}$, means that there is also a constant-velocity drift, with speed $v_{\parallel}$, taking place. This combination of linear and circular motion is called *helical motion* because the path of the particle is that of a helix.

Example 22–2 Motion in a Magnetic Field A uniform magnetic field of 3.75 T points in the positive x direction. Particles of mass 25.0 g and charge 6.20 mC are fired into this magnetic field at angles of 90.0°, 45.0°, and 0.00° above the positive x-axis, in the x-y plane, with a speed of 35.0 cm/s. Calculate the radius of each particle's path and describe its motion.

Picture the Problem The sketch shows the magnetic field vector together with three velocity vectors in the stated directions.

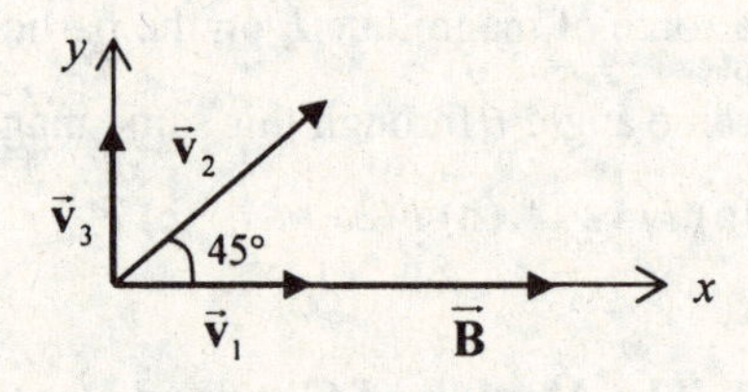

Strategy We can use the results discussed above to determine the radius in each case and then consider which type of motion we have.

Solution

1. For the case with $\vec{v}_1$, there is no component perpendicular to $\vec{B}$; therefore: we get constant-velocity motion with velocity $\vec{v}_1$.

2. For particle 2, the perpendicular component of $\vec{v}_2$ is given by: $v_{2\perp} = v_2 \sin(45^{\circ})$

3. The radius of the motion for particle 2 is: $r_2 = \frac{mv_{2\perp}}{qB} = \frac{(0.025\text{ kg})(0.350\frac{\text{m}}{\text{s}})\sin(45^{\circ})}{(6.20\times 10^{-3}\text{ C})(3.75\text{ T})} = 0.266\text{ m}$

4. The parallel component of $\vec{v}_2$ is: $v_{2\parallel} = v_2 \cos(45^{\circ}) = (0.350\text{ m/s})\cos(45^{\circ}) = 0.247\text{ m/s}$

5. The motion of particle 2 is: helical motion of radius 0.266 m with a constant speed drift of 0.247 m/s in the positive x direction

6. Particle 3 moves perpendicular to $\vec{\mathbf{B}}$. The radius of its path is: $r_3 = \frac{mv_3}{qB} = \frac{(0.025\text{ kg})(0.350\text{ m/s})}{(6.20\times10^{-3}\text{ C})(3.75\text{ T})} = 0.376\text{ m}$

7. Since $\vec{\mathbf{v}}_3$ has no parallel component: we have uniform circular motion with speed 35.0 cm/s and radius 37.6 cm.

Insight Each of the three types of motion is treated here.

Practice Quiz

5. For a particle of mass m and charge q moving at an angle θ with respect to the direction of a uniform magnetic field $\vec{\mathbf{B}}$, which of the following values of θ will produce the path of largest radius?

   (a) 0° **(b)** 30° **(c)** 45° **(d)** 60° **(e)** 90°

6. For a particle of mass m and charge q moving at an angle θ with respect to the direction of a uniform magnetic field $\vec{\mathbf{B}}$, which of the following values of θ will produce the fastest constant-speed drift along the direction of the magnetic field?

   (a) 0° **(b)** 30° **(c)** 45° **(d)** 60° **(e)** 90°

22–4 The Magnetic Force Exerted on a Current-Carrying Wire

Because a force is exerted on a moving charge in a magnetic field, and current consists of moving charges, a force will be exerted on a current-carrying wire in a magnetic field. The force exerted on this wire is the resultant of the forces on the moving charges that make up the current. The magnitude of this net result works out to be

$$F = ILB\sin\theta$$

where θ is the angle between the direction of current flow and the magnetic field. The direction of this force is given by the same magnetic force right-hand rule as for individual charges. In this case, the direction of the velocity (for a positive charge) is the same as the direction of the current.

Practice Quiz

7. A uniform magnetic field is directed into the page, ⊗. A current I flows vertically from the bottom of this page toward the top, ↑. What is the direction of the force on the wire carrying this current?

   (a) ↓ **(b)** → **(c)** ⊗ **(d)** ⊙ **(e)** None of the above.

8. A current I flows out of the page, ⊙. It sits in a magnetic field $\vec{\mathbf{B}}$ that's directed into the page, ⊗. What is the direction of the force on the wire carrying this current?

(a) ↓ **(b)** ← **(c)** ↑ **(d)** → **(e)** $F = 0$, so there is no direction.

22–5 Loops of Current and Magnetic Torque

When an entire loop of current sits in a magnetic field, the forces on the different parts of the loop can result in a net torque on the loop depending on how the loop is oriented with respect to the magnetic field. The magnitude of this torque is given by

$$\tau = NIAB\sin\theta$$

where N is the number of "turns," that is, the number of times the wire carrying the current I is wrapped around the loop, A is the area of the loop, and B is the magnitude of the magnetic field. In the case of the torque on a current loop, the angle θ refers to the angle between the magnetic field and the direction normal to the plane of the loop. The normal direction used to determine θ in the above expression is determined by its own right-hand rule

The normal direction for a current loop is in the direction of the thumb when the fingers of the right hand curl around the loop in the direction of the current.

The effect of this torque on a current loop is to rotate the loop such that its normal direction coincides with the direction of the magnetic field.

Example 22–3 Torque A conducting wire is wrapped once around a plastic rectangular slab that is 10.0 cm × 15.0 cm in size. The construction is attached to an axle so that it is free to rotate about its center. If a current of 1.50 A travels through the wire in a clockwise direction as viewed from the face of the slab, what is the magnitude and direction of a uniform magnetic field that causes the slab to experience a maximum torque of 2.50 N·m rotating it counterclockwise as viewed from above the right-hand side of the axle?

Picture the Problem The diagram shows the current loop and an eye indicating how the rotating loop is viewed.

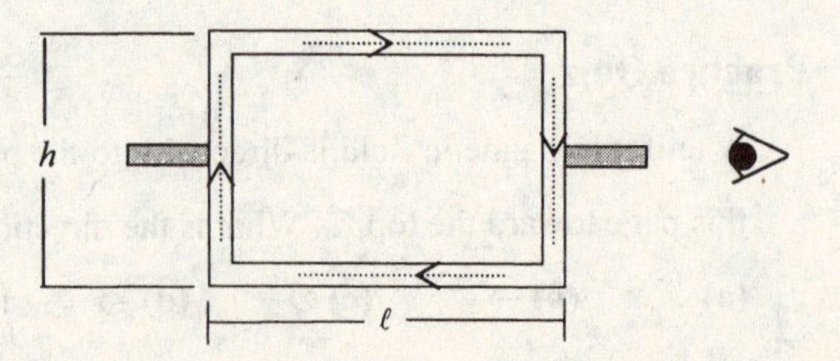

Strategy The maximum torque occurs for $\sin\theta = 1$, so we can apply the expression for torque to find

the magnitude of $\vec{\mathbf{B}}$. For the direction of $\vec{\mathbf{B}}$, we'll consider the needed direction of the force on a segment of wire.

Solution

1. Setting $\sin\theta = 1$, we can use the relationship between B and τ to solve for B:

$$\tau_{\max} = NIAB \Rightarrow B = \frac{\tau_{\max}}{NIA} = \frac{\tau_{\max}}{IA}$$

2. The area is $A = \ell h$ so B is:

$$B = \frac{\tau_{\max}}{I\ell h} = \frac{2.50\ \text{N}\cdot\text{m}}{(1.50\ \text{A})(0.150\ \text{m})(0.100\ \text{m})} = 111\ \text{T}$$

3. For the eye to see counterclockwise rotation: The direction of force on the upper wire is $\odot$.

4. Given the above direction of the force, the magnetic force right-hand rule gives a magnetic field direction: The direction of $\vec{\mathbf{B}}$ at the upper wire must be upward, $\uparrow$.

5. Thus, the magnetic field is: $\vec{\mathbf{B}}$ = 111 T, pointing upward.

Insight Verify that using this direction for $\vec{\mathbf{B}}$ on the other wire segments is consistent with the required torque.

Practice Quiz

9. A single turn, circular loop of current 2.3 A has a radius of 8.9 cm and experiences a maximum torque of 0.065 N·m when placed in a magnetic field. What is the magnitude of the magnetic field?
(a) 1.1 T **(b)** 1.3 T **(c)** 0.0 T **(d)** 0.32 T **(e)** 0.10 T

22–6 Electric Currents, Magnetic Fields, and Ampere's Law

The connection between current and magnetism goes further than just the force on a current-carrying wire. It turns out that currents are sources of magnetism in that a current creates a magnetic field in the space around it. Recall that magnetic fields form closed loops, so the magnetic field created by a current loops around the current. The direction in which the magnetic field loops around the current is determined by a **magnetic field right-hand rule**:

If you orient the thumb of your right hand along the current-carrying wire in the direction of the current, your fingers will curl around the wire in the direction of the magnetic field.

The closed loops formed by the magnetic field enclose the current that creates the field. The relationship between the enclosed current and the field it creates is given by **Ampere's law**:

$$\sum B_{\parallel}\Delta L = \mu_0 I_{enclosed}$$

In this expression, $B_{\parallel}$ is a component of $\overline{\mathbf{B}}$ that is parallel to a little segment, ΔL, of a path that encloses the current I_{enclosed}. The factor μ_0 is a fundamental constant called the **permeability of free space**; it has the value

$$\mu_0 = 4\pi \times 10^{-7}\ \text{T}\cdot\text{m/A}$$

Ampere's law is somewhat like Gauss's law in that it is most useful for finding the magnetic field in cases with a high degree of symmetry. One such case is that of a long, straight wire carrying a current I. In this case, Ampere's law gives the result for the magnitude of the magnetic field to be

$$B = \frac{\mu_0 I}{2\pi r}$$

Now that we know that a current both creates a magnetic field and experiences a force in a magnetic field, then two parallel current-carrying wires, I_1 and I_2, will exert forces on each other. Each current "feels" the force due to the magnetic field created by the other. They will exert equal and opposite forces on each other of magnitude

$$F = \frac{\mu_0 I_1 I_2}{2\pi d} L$$

where L is the length of the wires (assumed to be equal for simplicity). The direction of the force on each wire can be determined by the magnetic force right-hand rule. By applying this rule to each wire, we find that if the currents are in the same direction, the two wires attract each other, and if the currents are in opposite directions, the wires repel each other.

Example 22–4 Force between Currents and the Ampere The same current runs through two parallel wires that are separated by 1.00 m. If the force per unit length on each wire is 2.00×10^{-7} N/m, what current runs through the wires?

Picture the Problem The sketch shows two parallel wires carrying the same current I.

I

I

Strategy We can solve this problem by manipulating the equation for the force between the currents.

Solution

1. The force per unit length is: $F = \frac{\mu_0 I_1 I_2}{2\pi d} L \Rightarrow \left(\frac{F}{L}\right) = \frac{\mu_0 I^2}{2\pi d}$

2. The current is given by: $I = \sqrt{\frac{2\pi d (F/L)}{\mu_0}} = \sqrt{\frac{2\pi (1.00\text{ m})(2.00\times 10^{-7}\text{ N/m})}{4\pi\times 10^{-7}\text{ T}\cdot\text{m/A}}}$

$= 1.00\text{ A}$

Insight This calculation is the basis for using the ampere as the fundamental quantity of electricity rather than the coulomb. It turns out to be more accurate to measure one ampere of current this way than it is to measure one coulomb of charge.

Practice Quiz

10. A long, straight current I produces a magnetic field of magnitude B a distance r away from it. What is the magnitude of the magnetic field at a distance of $r/2$ from the wire?

    (a) $B/4$ **(b)** $B/2$ **(c)** B **(d)** $2B$ **(e)** $4B$

11. Two parallel wires of equal length, separated by a distance d, carry the same current I in opposite directions. If the current is reduced to $I/2$, the force between the wires becomes

    (a) $I/4$. **(b)** $I/2$. **(c)** I. **(d)** $2I$. **(e)** $4I$.

22–7 Current Loops and Solenoids

In the previous section, the result for the magnetic field of a long, straight wire was given. There are other useful configurations of current-carrying wires such as current loops and **solenoids**. For a circular loop of wire of radius R, carrying a current I, and containing N turns, the magnitude of the magnetic field at the center of the loop is given by

$$B = \frac{N\mu_0 I}{2R}$$

A solenoid is an electrical device in which a wire has been wound into the geometry of a helix. The circular loops are usually tightly packed. Solenoids are sometimes referred to as *electromagnets*. A solenoid carrying a current I produces a nearly uniform magnetic field inside the loops. Ampere's law shows that the magnitude of this field is given by

$$B = \mu_0 \frac{N}{L} I = \mu_0 n I$$

where N is the number of turns and L is the length of the solenoid. The quantity $n = N/L$ is the number of turns per unit length.

Practice Quiz

12. How many turns per centimeter are needed to produce a 1.00-T magnetic field inside a solenoid with a 10.0-A current?

(a) 10.0 **(b)** 796 **(c)** 1.00 **(d)** 79,600 **(e)** $4\pi \times 10^7$

22–8 Magnetism in Matter

The magnetic properties of matter is an important branch of applied physics. The very fact that materials have magnetic properties can be traced back to the fact that the electrons in the atoms of a substance possess little magnetic fields as part of their fundamental character. For many atoms, there is a net cancellation of these little magnetic fields. However, the structure of some substances (as with iron and nickel) is such that there is a nonzero net magnetic field. In bulk, the magnet fields of these atoms tend to align, giving rise to a permanent magnetic field. Materials that behave this way are called **ferromagnets.** Ferromagnetic materials can often lose their magnetism at high temperatures.

In some materials, the tendency for the magnetic fields of the atoms to align themselves is too weak to produce a significant effect. However, an external magnetic field applied to the material will produce alignment, resulting in a magnetized material. This behavior is called **paramagnetism**. We briefly mention one more effect, called **diamagnetism**, in which an external magnetic field applied in a given direction produces an oppositely directed magnetic field in the material.

Reference Tools and Resources

I. Key Terms and Phrases

north pole the north pole of a magnet is the end that tends to point northward with respect to the earth

south pole the south pole of a magnet is the end that tends to point southward with respect to the earth

magnetic field the field created by a magnetic material or a current that carries its magnetic effect

magnetic field lines a pictorial representation of a magnetic field

tesla the SI unit of the magnetic field

magnetic force right-hand rule the rule for determining the direction of the magnetic force on a moving charge or current. For a positive charge, place the fingers of your right hand along $\vec{v}$ so that they would curl toward $\vec{B}$; your thumb then gives the direction of $\vec{F}$. Reverse the direction for a negative charge

magnetic field right-hand rule the rule for determining the direction of the magnetic field created by a current. Orient the thumb of your right hand along the current; your fingers then curl in the direction of $\vec{B}$

Ampere's law the fundamental law relating a current to the magnetic field it creates

permeability of free space the fundamental constant of magnetism in Ampere's law

solenoid an electrical device in which a wire has been wound into the geometry of a helix

ferromagnetism the phenomenon that some materials have a spontaneously created "permanent" magnetic field

paramagnetism the phenomenon that some materials produce a magnetic field in the same direction as an external magnetic field

diamagnetism the phenomenon that matter produces a magnetic field in the opposite direction of an external magnetic field

II. Important Equations

Name/Topic	Equation	Explanation
magnetic force on a moving charge	$F = qvB\sin\theta$	The force law for the magnitude of the force on a moving charge in a magnetic field
magnetic force on a current-carrying wire	$F = ILB\sin\theta$	The magnitude of the magnetic force on a current-carrying wire
electric currents and magnetic fields	$B = \dfrac{\mu_0 I}{2\pi r}$	The magnitude of the magnetic field due to a long, straight current I
two parallel current-carrying wires	$F = \dfrac{\mu_0 I_1 I_2}{2\pi d} L$	The magnitude of the force between two parallel current-carrying wires

solenoids	$B = \mu_0 \frac{N}{L} I = \mu_0 n I$	The magnitude of the magnetic field inside a solenoid

III. Know Your Units

Quantity	Dimension	SI Unit
magnetic field ($\vec{\mathbf{B}}$)	$[M][A^{-1}][T^{-2}]$	T
permeability of free space (μ_0)	$[M][L][A^{-2}][T^{-2}]$	T·m/A

Puzzle

THE SNAKE

The figure shows a flexible wire loop lying on a frictionless table. The wire can slide around and change its shape freely, but it cannot move perpendicular to the table.

Part a

If a current were made to flow counterclockwise in the loop, which of the following would happen?

1. The wire would stretch out into a circle.
2. The wire would contract down to a tangled mess.
3. Nothing would happen.
4. There is not enough information to tell.

Part b

If the current went clockwise, would that affect your answer to part a? Answer this in words, not equations, briefly explaining how you obtained your answer.

Answers to Selected Conceptual Questions and Exercises

Conceptual Questions

2. Yes. If an electric field exists in this region of space, and no magnetic field is present, the electric field will exert a force on the electron and cause it to accelerate.

4. In a uniform electric field, the force on a charged particle is always in one direction, as with gravity near the Earth's surface, leading to parabolic trajectories. In a uniform magnetic field the force of a charged particle is always at right angles to the motion, resulting in circular or helical trajectories.

Perhaps even more important, a charged particle experiences a force due to an electric field whether it is moving or at rest; in a magnetic field, the particle must be moving to experience a force.

6. A current-carrying wire in a uniform magnetic field can experience zero force only if the wire points in the same or opposite direction as the magnetic field. In such a case, the angle θ in Equation 22-4 will be either 0° or 180°, in which case $F = ILB \sin\theta = 0$.

Conceptual Exercises

10. **(a)** Note that the direction of the magnetic force is given by the right-hand rule, and that the direction of the electric force is in the direction of the electric field. Applying these considerations, we find that the ranking in order of increasing magnitude of the net force is as follows: $F_3 < F_2 = F_4 < F_1$. **(b)** The only velocity that could result in zero net force is $\vec{v}_3$, since it is the only velocity that produces a magnetic force that is opposite in direction to the electric force.

16. A simple sketch of the fields produced at the center of the triangle by the three wires shows that the net field points in a direction between $\theta = 0$ and $\theta = 90^\circ$ where θ is measured relative to the positive x axis. More careful analysis shows that the net field points at an angle of $\theta = 60^\circ$ relative to the positive x axis, and has a magnitude equal to twice the magnitude of the field produced by one wire alone.

20. If the current loop is to attract the magnet, it must produce a magnetic field with its north pole pointing to the right; that is, pointing toward the south pole of the bar magnet. For this to be the case, the current in the wire must flow from terminal A to terminal B, which means, in turn, that terminal A must be the positive terminal.

Solutions to Selected End-of-Chapter Problems

17. **Picture the Problem:** An electron moves into a constant magnetic field oriented perpendicular to its path. It then follows a circular trajectory.

Strategy: The electron's energy is conserved. This fact will determine its speed.

Solution:

1. Apply the conservation of energy for the electron and solve for the speed: $\frac{1}{2}mv^2 = eV \Rightarrow v = \sqrt{\frac{2eV}{m}}$

2. Rewrite the expression for the radius of the path and solve for B: $r = \frac{mv}{eB} \Rightarrow B = \frac{mv}{er}$

3. Insert the expression for v and solve:

$$B = \frac{m}{er}\sqrt{\frac{2eV}{m}} = \frac{1}{r}\sqrt{\frac{2mV}{e}}$$

$$= \frac{1}{0.17\text{ m}}\sqrt{\frac{2(9.11\times10^{-31}\text{ kg})(410\text{ V})}{1.60\times10^{-19}\text{ C}}} = \boxed{0.40\text{ mT}}$$

Insight Never forget the importance and usefulness of the conservation of energy.

19. Picture the Problem: A charged particle follows a circular path in a magnetic field.

Strategy: Use the right-hand-rule to determine the charge of the particle. The expression for the radius involves the mass.

Solution:

Part (a)

1. By the right-hand rule, a positive charge would feel a force to the left. Since the particle is experiencing a force to the right, it must have $\boxed{\text{negative}}$ charge.

Part (b)

2. Rewrite the expression for the radius of the path to solve for m:

$$r = \frac{mv}{eB} \quad \Rightarrow \quad m = \frac{erB}{v}$$

3. Insert the numerical values for the final result:

$$m = \frac{(1.60\times10^{-19}\text{ C})(0.520\text{ m})(0.180\text{ T})}{\left(1.67\times10^{-27}\ \frac{\text{kg}}{\text{u}}\right)\left(6.0\times10^{6}\ \frac{\text{m}}{\text{s}}\right)} = \boxed{1.5\text{ u}}$$

Insight Keep in mind that negative charges move oppositely from what the right-hand-rule gives.

35. Picture the Problem: A current loop is free to rotate in a region of uniform magnetic field.

Strategy: We can apply the expression for the torque on a current loop.

Solution:

1. The torque on a loop of current is: $\tau = IAB\sin\theta$

2. Since we are given that $\tau = \frac{1}{2}\tau_{\text{max}}$: $\sin\theta = 1/2$

3. Solving for θ gives the final result: $\theta = \sin^{-1}(1/2) = \boxed{30°}$

Insight Don't forget about the angular dependence of the torque on a loop of current.

47. Picture the Problem: Two long, straight wires are oriented perpendicular to the page.

Strategy: The magnetic field at P is the sum of the fields produced by the two wires. Let I_1 be located at the origin and I_2 be located on the x-axis. Then the coordinates of P are (0, 5.0 cm).

Solution:

1. Sum the magnetic fields from I_1 and I_2: $\vec{\mathbf{B}} = \vec{\mathbf{B}}_1 + \vec{\mathbf{B}}_2 = \dfrac{\mu_0 I_1}{2\pi r_1}\hat{\mathbf{x}} + \dfrac{\mu_0 I_2}{2\pi r_2}(\hat{\mathbf{x}}\cos\theta + \hat{\mathbf{y}}\sin\theta)$

$$= \frac{\left(4\pi\times10^{-7}\ \frac{\text{T}\cdot\text{m}}{\text{A}}\right)}{2\pi}\left[\frac{3.0\ \text{A}}{0.050\ \text{m}}\hat{\mathbf{x}} + \frac{4.0\ \text{A}}{\sqrt{2(0.050\ \text{m})^2}}(\hat{\mathbf{x}}\cos 225° + \hat{\mathbf{y}}\sin 225°)\right]$$

$$= (4.0\times10^{-6}\ \text{T})\hat{\mathbf{x}} - (8.0\times10^{-6}\ \text{T})\hat{\mathbf{y}}$$

2. The magnitude of $\vec{\mathbf{B}}$ then is: $B = \sqrt{(4.0\times10^{-6}\ \text{T})^2 + (-8.0\times10^{-6}\ \text{T})^2} = 8.9\times10^{-6}\ \text{T}$

3. The direction of $\vec{\mathbf{B}}$ is: $\theta = \tan^{-1}\dfrac{-8.0\times10^{-6}\ \text{T}}{4.0\times10^{-6}\ \text{T}} = 63°$ (below the dashed line to the right)

Insight Make sure you see why $\theta = 225°$.

67. Picture the Problem: Two parallel wires carry equal currents in the same direction.

Strategy: The magnetic field at each point is the vector sum of the magnetic fields due to each wire.

Solution:

1. Sum the magnetic fields from each wire at point A: $\vec{\mathbf{B}}_A = \vec{\mathbf{B}}_{1A} + \vec{\mathbf{B}}_{2A} = \dfrac{\mu_0 I}{2\pi r_1}\odot + \dfrac{\mu_0 I}{2\pi r_2}\odot = \dfrac{\mu_0 I}{2\pi}\left(\dfrac{1}{r_1} + \dfrac{1}{r_2}\right)\odot$

$$= \frac{\left(4\pi\times10^{-7}\ \frac{\text{T}\cdot\text{m}}{\text{A}}\right)(2.2\ \text{A})}{2\pi}\left(\frac{1}{0.075\ \text{m}} + \frac{1}{3(0.075\ \text{m})}\right)\odot$$

$$= \boxed{(7.8\ \mu\text{T})\odot}$$

2. Sum the magnetic fields from each wire at point B: $\vec{\mathbf{B}}_B = \vec{\mathbf{B}}_{1B} + \vec{\mathbf{B}}_{2B} = \dfrac{\mu_0 I}{2\pi}\left(\dfrac{1}{r_1}\otimes + \dfrac{1}{r_2}\odot\right)$

$$= \frac{\mu_0 I}{2\pi}\left(\frac{1}{r} - \frac{1}{r}\right)\odot = \boxed{0}$$

3. Insert the expression for v and solve: $\vec{\mathbf{B}}_C = \vec{\mathbf{B}}_{1C} + \vec{\mathbf{B}}_{2C} = \dfrac{\mu_0 I}{2\pi}\left(\dfrac{1}{r_1}\otimes + \dfrac{1}{r_2}\otimes\right)$

$$= \frac{\left(4\pi\times10^{-7}\ \frac{\text{T}\cdot\text{m}}{\text{A}}\right)(2.2\ \text{A})}{2\pi}\left(\frac{1}{3(0.075\ \text{m})} + \frac{1}{0.075\ \text{m}}\right)\otimes$$

$$= \boxed{(7.8\ \mu\text{T})\otimes}$$

Insight Once you understand how to get the magnitude of the magnetic field it becomes vector addition in one dimension.

Answers to Practice Quiz

1. (c) **2.** (d) **3.** (b) **4.** (b) **5.** (e) **6.** (a) **7.** (e) **8.** (e) **9.** (a) **10.** (d) **11.** (a) **12.** (b)

CHAPTER 23

MAGNETIC FLUX AND FARADAY'S LAW OF INDUCTION

Chapter Objectives

After studying this chapter, you should

1. know how to calculate the magnetic flux through a surface of area *A*.
2. know the relationship between induced emf and magnetic flux: Faraday's law of induction.
3. be able to use Lenz's law to determine the direction of an induced current.
4. know the relationship between a magnetic field and an induced electric field.
5. understand the basic principle of how electric generators produce alternating current.
6. understand the concept of self-inductance and the behavior of *RL* circuits.
7. be able to calculate the energy stored in the magnetic field of an inductor.
8. understand how step-up and step-down transformers work.

Warm-Ups

1. Lenz's law is said to be a consequence of the principle of conservation of energy. Explain this statement by describing what would happen if Lenz's law were reversed. Take the example of a wire loop with a changing magnetic flux, and describe what would happen if current were induced opposite to the way it really is.
2. Let's say you take an ordinary wire coat hanger and straighten out the hook shaped part. Now, you can spin the (roughly) triangular part around by twisting the straightened part between your fingers. Estimate the EMF that you can generate by spinning the hanger in Earth's magnetic field (about 5×10^{-5} T).

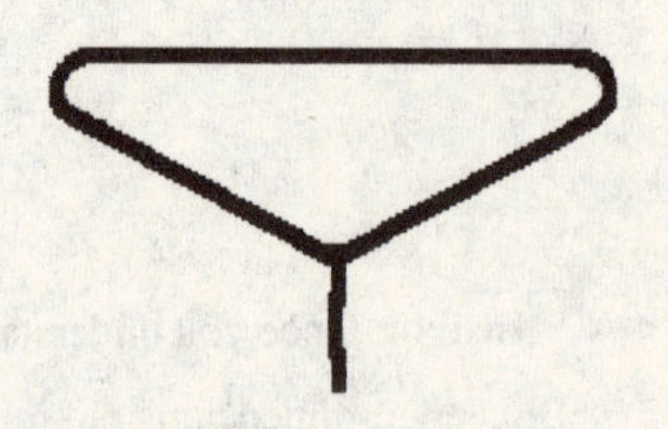

3. Here's one way of understanding a capacitor: It is a device that won't let the voltage between two points change too rapidly, because it stores up charge and has $V = Q/C$. The charge cannot be changed instantaneously, so the voltage cannot either. Describe an inductor in a similar way, that is, say what cannot be changed rapidly, and why.

4. Estimate the inductance of a solenoid made by buying a typical spool of wire from a hardware store and winding it carefully around a broom handle.
5. A wire hoop surrounds a long solenoid. If the current in the solenoid increases, will there be any current induced in the loop? If so, which way will the current flow?
6. A car battery has an emf of only 12 V, yet energy from the battery provides the 20,000-V spark that ignites the gasoline. How is this possible?

Chapter Review

This chapter provides more detail about the connection between electricity and magnetism. The very important concept of electromagnetic induction is introduced. Some of the most important applications of this concept, such as electric generators, motors, and transformers, are also discussed.

23–1 – 23–2 Induced Electromotive Force and Magnetic Flux

The fundamental fact on which this chapter is based is that a magnetic field, when it changes, can produce an electric field. This is seen in the fact that when the magnetic field passing through a coil changes, an **induced current** is observed in the coil. The induced current is present just as if a battery had been placed in the coil, so we say that there is an **induced emf** as a result of the changing magnetic field. The magnitude of the induced emf is directly proportional to the rate at which the magnetic field changes.

The precise magnitude of the induced emf is well represented in terms of the **magnetic flux** Φ through the coil. As with electric flux, magnetic flux represents the "flow" of the magnetic field through a surface. The magnetic flux through a surface of area A is calculated in the same way as the electric flux:

$$\Phi = BA\cos\theta$$

where again θ is the angle between the magnetic field and the direction normal to the surface. The SI unit of magnetic flux is called the **weber** (Wb): 1 Wb = 1 T·m^2.

Example 23–1 Magnetic Flux A uniform magnetic field of magnitude 0.250 T makes an angle of 35° with the normal to a circular aperture of radius 65 cm. If the aperture lies entirely in the magnetic field, what is the magnetic flux through the aperture?

Picture the Problem The sketch shows a circular aperture, the magnetic field vector and a unit vector, $\hat{\mathbf{n}}$ representing the direction normal to the surface.

n̂
35°
$\vec{B}$

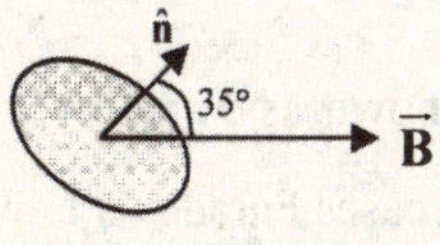

Strategy The expression for the magnetic flux through an area A can be used directly for this calculation. We need only to first calculate the area of the aperture.

Solution

1. The area of the circular aperture is: $A = \pi r^2$

2. The magnetic flux then is: $\Phi = B(\pi r^2)\cos(\theta) = (0.250\ \text{T})\pi(0.65\ \text{m})^2\cos(35°)$

$= 0.27\ \text{Wb}$

Insight This example is almost the same as Example 19–4 for electric flux. Thus, if you understood electric flux then, you should understand magnetic flux now.

Practice Quiz

1. What angle between $\vec{\mathbf{B}}$ and the normal to a surface produces the largest magnetic flux through that surface?

   (a) 0° **(b)** 30° **(c)** 45° **(d)** 60° **(e)** 90°

2. A magnetic field $\vec{\mathbf{B}}$ passes through an area A, at angle θ, giving rise to a magnetic flux Φ. If B is doubled and A is halved, the magnetic flux through the area will be

   (a) $\Phi/4$. **(b)** $\Phi/2$. **(c)** Φ. **(d)** 2Φ. **(e)** 4Φ.

23–3 – 23–4 Faraday's Law of Induction and Lenz's Law

Using the concept of magnetic flux, we can write an expression to determine the precise magnitude of the induced emf that arises in a coil when the magnetic field changes. What is observed is that the magnitude of the induced emf $\mathcal{E}$ in a single coil equals the rate of change of the magnetic flux through the coil. If the coil has N turns, the result for the induced emf is

$$\mathcal{E} = -N\frac{\Delta\Phi}{\Delta t}$$

This result is known as **Faraday's law of induction**. Often, only the magnitude of the induced emf is desired; in such cases Faraday's law is simply written as $|\mathcal{E}| = N|\Delta\Phi/\Delta t|$.

The minus sign in Faraday's law is there to account for the polarity (or direction) of the induced emf. The polarity of an induced emf is correctly determined by applying **Lenz's law**:

> *The polarity of the induced emf in a coil is such that the induced current creates a magnetic field whose flux opposes the original change in flux.*

Notice, the law says only that the flux due to the induced magnetic field *opposes* the original change in flux, but it does not cancel the original change in flux. As explained in the text, this result is the only possibility that preserves the conservation of energy.

Example 23–2 Faraday's Law A square loop of side length 25.0 cm sits in a magnetic field that makes an angle of 60.0° with the normal to the loop. If the magnetic field increases uniformly from 0.250 T to 1.25 T in 0.525 seconds, determine the magnitude and polarity of the induced emf in the loop.

Picture the Problem The sketch shows the square loop, the magnetic field vector, and a unit vector $\hat{\mathbf{n}}$ representing the direction normal to the surface.

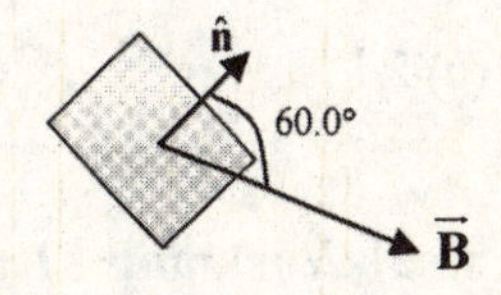

Strategy Using the expression for the area of a square, we can get the change in flux using the expression for magnetic flux. Direct application of Faraday's and Lenz's laws then give the final results.

Solution

1. If we call the side length d, the area is: $A = d^2$

2. The change in flux is:
$$\Delta\Phi = \Phi_f - \Phi_i = B_f d^2 \cos\theta - B_i d^2 \cos\theta = d^2 \cos\theta\,(B_f - B_i)$$

3. Faraday's law for the magnitude of the induced emf gives:
$$|\mathcal{E}| = d^2\cos\theta(B_f - B_i)/\Delta t = (0.250\text{ m})^2\cos(60.0^\circ)\frac{[1.25\,T - 0.250\text{ T}]}{0.525\text{ s}} = 0.0595\text{ V}$$

4. In the diagram, the flux due to $\vec{\mathbf{B}}$ is increasing outward, so: We must have an inward induced magnetic flux.

5. For an inward induced flux, the magnetic field a clockwise-induced current and therefore a right-hand rule gives: clockwise polarity for the induced emf.

Insight Reconstruct the Lenz's law argument if $\vec{\mathbf{B}}$ had been decreasing instead of increasing.

Practice Quiz

3. For case 1, the current in a coil is constant, for case 2, the current in the same coil changes at a rate of +1 A/s, and for case 3, the current in this coil changes at a rate of –2 A/s. For which case will there be the largest induced emf?

   (a) case 1 **(b)** case 2 **(c)** case 3 **(d)** It cannot be determined from the given information.

4. If each of the arrows in the figure on the right represents an increasing current, which one will produce a clockwise-induced emf in the circular coil?

   (a) 1 **(b)** 2 **(c)** 3 **(d)** 4 **(e)** None

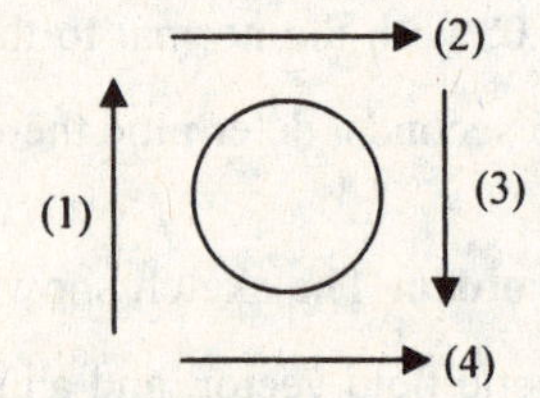

23–5 Mechanical Work and Electrical Energy

The magnetic flux through a loop can change either as a result of changing the magnetic field that passes through the loop, changing the orientation of the loop within the field, or by changing the area of the loop. One way that the area of the loop can change is by moving the conducting wire that makes up the loop; this latter possibility leads directly to the concept of **motional emf**. A motional emf is an induced emf that results from the motion of a conductor through a magnetic field $\vec{\mathbf{B}}$. Applying Faraday's law to this situation shows that the magnitude of the induced emf is given by

$$|\mathcal{E}| = Bv\ell$$

where v is the speed of the conductor, and ℓ is its length. To be a little more precise, in the above expression, ℓ, $\vec{\mathbf{v}}$, and $\vec{\mathbf{B}}$ are mutually perpendicular.

An emf, induced or otherwise, is best understood as coming from an electric field. In fact, we know that a changing magnetic flux creates an electric field. The result from the induced motional emf helps us to see the relationship between the magnetic field through which the conductor moves and the electric field induced by the changing flux:

$$E = Bv$$

In this simple case, both E and B are constant. When we study light (or electromagnetic radiation) a relationship just like the above expression will be very important even though both the electric and magnetic fields will be changing.

If a motional emf is established in a circuit of resistance R, a current $I = Bv\ell / R$ will flow through this circuit. The magnetic force on this current will act to stop the motion of this conductor unless an equal and opposite external force is applied to maintain the constant speed v. The mechanical power supplied by this external force is precisely what supplies the electrical power consumed through the resistor at any moment. This power is

$$P_{mechanical} = P_{electrical} = \frac{(Bv\ell)^2}{R}$$

Exercise 23–3 An Induced Electric Field A metal bar 1.5 m in length moves in a direction perpendicular to a uniform magnetic field of magnitude 0.85 T with a speed of 2.2 m/s. Determine the magnitude of the induced electric field in the metal bar.

Solution: We are given the following information:

Given: $\ell = 1.5$ m, $B = 0.85$ T, $v = 2.2$ m/s; **Find:** E

The relationship between the electric and magnetic fields is $E = Bv$. Therefore,

$$E = (0.85 \text{ T})(2.2 \text{ m/s}) = 1.9 \text{ N/C}$$

Insight You should verify that the units work out correctly. As mentioned above, the fact that induction relates E and B through a velocity is important in understanding electromagnetic radiation.

Practice Quiz

5. A conductor of length L moves with speed v through a magnetic field $\vec{\mathbf{B}}$, producing a motional emf $\mathcal{E}$. If the length of the conductor is doubled, the induced emf will be

   (a) $\mathcal{E}/4$. **(b)** $\mathcal{E}/2$. **(c)** $\mathcal{E}$. **(d)** $2\mathcal{E}$. **(e)** $4\mathcal{E}$.

6. A conductor of length L moves with speed v through a magnetic field $\vec{\mathbf{B}}$, producing an induced electric field E. If the length of the conductor is doubled, the induced electric field will be

   (a) $E/4$. **(b)** $E/2$. **(c)** E. **(d)** $2E$. **(e)** $4E$.

23–6 Generators and Motors

Faraday's law of induction has had a tremendous impact on society. This law is the basic principle behind **electric generators**. An electric generator is a device designed to convert mechanical energy to electrical energy. To accomplish this conversion, most modern generators rotate a loop of area A and N turns at an angular speed ω in a magnetic field $\vec{\mathbf{B}}$. This rotation causes a continuous change in the magnetic flux through the loop that results in an induced emf given by

$$\mathcal{E} = NBA\omega \sin \omega t$$

Notice that as time progresses the induced emf will change sign. When the induced emf changes its sign, the induced current changes direction and thus is an **alternating current** (AC). Generators of this type are called AC generators for this reason.

An **electric motor** has the opposite purpose of an electric generator. Motors are designed to convert electrical energy into mechanical work. The basic physics behind the workings of a motor is the torque on a current loop discussed in Chapter 22. Electrical power supplies current to a loop that sits in a magnetic field. This loop then experiences a torque, which causes the loop to rotate. The rotation of this loop can then be used to do mechanical work such as the turning of a wheel.

Practice Quiz

7. A coil of 2000 turns and area 1.5 m^2 rotates in a magnetic field of 10 T with an angular velocity of 377 rad/s. What is the maximum emf generated in this coil?

   (a) 11 MV **(b)** 1.4 MV **(c)** 0 V **(d)** 2000 V **(e)** 20 kV

23–7 Inductance

When a current flows through a loop, the magnetic field created by that current has a magnetic flux through the area of the loop. If the current changes, the magnetic field changes, and so the flux changes, giving rise to an induced emf. This phenomenon is called **self-induction** because it is the loop's own current, and not an external one, that gives rise to the induced emf. Because $\Delta\Phi/\Delta t \propto \Delta B/\Delta t \propto \Delta I/\Delta t$, Faraday's law implies that $\mathcal{E} \propto \Delta I/\Delta t$. The constant of proportionality, L, is called the **inductance** of the loop,

$$\mathcal{E} = -L\frac{\Delta I}{\Delta t}$$

The SI unit of inductance is the **henry** (H), where 1 H = 1 V·s/A.

Comparing the self-inductance form of Faraday's law with the original form, $\mathcal{E} = -N(\Delta\Phi/\Delta t)$, we see a useful way to calculate the inductance of a coil, $L = N(\Delta\Phi/\Delta I)$. Using this result shows that for a solenoid

$$L = \mu_0\left(\frac{N^2}{\ell}\right)A = \mu_0 n^2 A\ell$$

where $n = N/\ell$ is the number of turns per unit length of the solenoid.

Example 23–4 Self-Inductance in a Solenoid A solenoid of length 12.5 cm contains 250 turns and has a radius of 1.75 cm. If the current in the solenoid decreases from 1.5 A to 0.45 A in 0.0125 seconds, what is the magnitude of the induced emf in the solenoid?

Picture the Problem The sketch shows a coil representing the solenoid in a circuit with an emf.

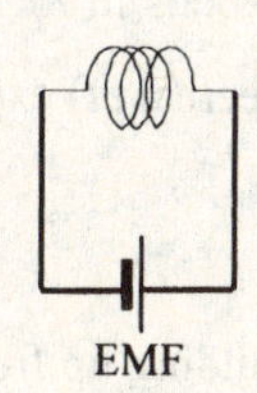

Strategy We can solve this problem using the self-inductance form of Faraday's law, but we first need to get the inductance of the solenoid.

Solution

1. The inductance of the solenoid is:

$$L = \mu_0\left(\frac{N^2}{\ell}\right)A = \mu_0\left(\frac{N^2}{\ell}\right)\pi r^2$$

$$= \left(4\pi\times10^{-7}\ \tfrac{\text{T}\cdot\text{m}}{\text{A}}\right)\left[\frac{(250)^2}{0.125\ \text{m}}\right]\pi(0.0175\ \text{m})^2 = 0.6045\ \text{mH}$$

2. Faraday's law gives:

$$|\mathcal{E}| = L\frac{|\Delta I|}{\Delta t} = L\frac{|I_f - I_i|}{\Delta t}$$

$$= \left(0.6045\times10^{-3}\ \text{H}\right)\frac{|0.45\ \text{A} - 1.5\ \text{A}|}{0.0125\ \text{s}} = 0.051\ \text{V}$$

Insight Make sure you understand the absence of the minus sign and the use of absolute values in this calculation.

Practice Quiz

8. A solenoid of length ℓ and area A consists of N turns producing an inductance L. If the number of turns is doubled, the inductance will be

   (a) $L/4$. **(b)** $L/2$. **(c)** L. **(d)** $2L$. **(e)** $4L$.

23–8 RL Circuits

A coil with a finite inductance (and negligible resistance) is called an **inductor**. Inductors are often used as circuit elements; a circuit with a resistor and an inductor in series is called an ***RL* circuit**. Recall that induction in a coil tends to resist changes in the current. As a result, the current in a circuit that contains an inductor does not rise and fall as quickly as it otherwise would; the inductance causes it to rise and fall gradually. The larger the inductance, the more gradual is the change in current.

Similar to what happens in *RC* circuits, when an emf is first applied (or shut off) in an *RL* circuit, the current increases (or decreases) exponentially with a characteristic time given by the time constant

$$\tau = \frac{L}{R}$$

For a current that is building up from zero in an *RL* circuit, the result is

$$I = \frac{\mathcal{E}}{R}\left(1 - e^{-t/\tau}\right) \qquad \text{[building up]}$$

For a current that is falling off from the maximum value, $I_{max} = \mathcal{E}/R$, the result is

$$I = \frac{\mathcal{E}}{R}e^{-t/\tau} \qquad \text{[falling off]}$$

We can see that in the case where the current is falling off, τ repesents the amount of time it takes for the current to fall to $(e^{-1})I_{max} = (0.368)I_{max}$, that is, to 36.8% of its maximum value.

Practice Quiz

9. An *RL* circuit contains an inductor of inductance L and a resistor of resistance R giving a time constant of value τ. If the resistance is doubled, the time constant will be

   (a) $\tau/4$. **(b)** $\tau/2$. **(c)** τ. **(d)** 2τ. **(e)** 4τ.

10. Which of the following statements is true concerning the current in an *RL* circuit?

    (a) The time constant is how long it takes the current to build up to 36.8% of its maximum value.

    (b) The time constant is how long it takes the current to build up to 63.2% of its maximum value.

    (c) The time constant is how long it takes the current to fall off to 63.2% of its maximum value.

(d) The time constant is how long it takes the current to fall to 36.8% of its maximum value from 63.2% of it.

(e) The time constant is how long it takes the current to build up to 63.2% of its maximum value from 36.8% of it.

23–9 Energy Stored in a Magnetic Field

Because the induced current in an inductor resists the buildup of current in the coil, it therefore requires energy to build up the current in an inductor against this "resistance." Once the current is established, so is the magnetic field within the coil. The energy needed to build up this current is stored in the magnetic field. For an inductor of inductance L, sustaining a current I, the amount of energy U stored in the magnetic field is

$$U = \frac{1}{2}LI^2$$

Just as with the energy stored in the electric field of a capacitor, the relationship between the energy stored in an inductor and its magnetic field is nicely expressed in terms of the energy density (energy per unit volume) of the magnetic field

$$\text{energy density} = u_B = \frac{B^2}{2\mu_0}$$

This expression for the energy density applies to any magnetic field, not just that for an inductor.

Practice Quiz

11. A solenoid of inductance L sustains a current I storing an amount of energy U. If the current in the solenoid is reduced to half its previous value, the energy stored in the solenoid will be
    (a) $U/4$. **(b)** $U/2$. **(c)** U. **(d)** $2U$. **(e)** $4U$.

23–10 Transformers

A very important practical device that relies on the phenomenon of induction is the **transformer**. A transformer is a device that uses induction to increase or decrease the potential difference in a circuit. A transformer consists of a *primary coil*, containing N_p turns, across which an AC voltage V_p is applied. The primary coil is bound by an iron core to a secondary coil, containing N_s turns. By induction there will be an induced potential difference V_s across the secondary coil. The relationship between the primary and secondary voltages is given by the **transformer equation**:

$$\frac{V_p}{V_s} = \frac{N_p}{N_s}$$

When $V_s > V_p$ we have a **step-up transformer** and when $V_s < V_p$ we have a **step-down transformer.**

By conservation of energy, the average power in the secondary circuit must equal the average power in the primary circuit, $I_s V_s = I_p V_p$. Therefore,

$$\frac{I_s}{I_p} = \frac{V_p}{V_s} = \frac{N_p}{N_s}$$

so that a step-up transformer steps down the current and vice versa.

Exercise 23–5 A Big Step Down Suppose you want to use one transformer to step down from a power line voltage of 50,000 V to a household voltage of 120 V. If your primary coil consists of 10,000 turns, how many turns will you need in your secondary coil?

Solution: We are given the following information:

Given: $V_p = 50{,}000$ V, $V_s = 120$ V, $N_p = 10{,}000$; **Find:** N_s

The transformer equation gives

$$\frac{V_p}{V_s} = \frac{N_p}{N_s} \quad \Rightarrow \quad N_s = N_p\left(\frac{V_s}{V_p}\right) = 10{,}000\left(\frac{120\text{ V}}{50{,}000\text{ V}}\right) = 24$$

Insight In practice, the voltage from power lines is stepped down in several stages, not all at once.

Practice Quiz

12. The primary coil of a transformer contains 100 turns, and the secondary coil contains 200 turns. If the current in the primary coil is I, what is the current in the secondary coil?

(a) I (b) $2I$ (c) $I/2$ (d) $I/4$ (e) $4I$

Reference Tools and Resources

I. Key Terms and Phrases

induced emf a potential difference created by a changing magnetic flux

magnetic flux a measure of the extent to which a magnetic field "flows" through an area

weber the SI unit of magnetic flux, equal to 1 $\text{T}\cdot\text{m}^2$

Faraday's law of induction the induced emf in a coil equals the rate of change of magnetic flux through the coil

Lenz's law the polarity of an induced emf is such that the induced current gives rise to a magnetic flux that opposes the original change in flux

motional emf an induced emf that results from the motion of a conductor through a magnetic field

electric generator a device designed to convert mechanical energy to electrical energy

alternating current (AC) current that alternates in direction

electric motor a device designed to convert electrical energy into mechanical work

self-induction the creation of an induced emf in a coil due to its own changing current

inductance the proportionality constant between the induced emf in a coil and the rate of change in the current in that same coil

henry the SI unit of inductance

inductor a coil with a finite inductance used as a circuit element

transformer uses induction to increase (step-up) or decrease (step-down) the voltage in a circuit

II. Important Equations

Name/Topic	Equation	Explanation
magnetic flux	$\Phi = BA\cos\theta$	Magnetic flux measures the amount of magnetic field "flowing" through an area
Faraday's law	$\mathcal{E} = -N\dfrac{\Delta\Phi}{\Delta t}$	The induced emf from the rate of change in magnetic flux through a coil of N turns
motional emf	$\lvert\mathcal{E}\rvert = Bv\ell$	The magnitude of the induced emf from the motion of a conductor through a magnetic field
electric generators	$\mathcal{E} = NBA\omega\sin\omega t$	The emf of an AC generator

inductance	$\mathcal{E} = -L\dfrac{\Delta I}{\Delta t}$	Faraday's law in terms of self-inductance
RL circuits	$\tau = \dfrac{L}{R}$	The time constant for an *RL* circuit
stored energy	$U = \dfrac{1}{2}LI^2$	The energy stored in the magnetic field of an inductor
transformers	$\dfrac{V_p}{V_s} = \dfrac{N_p}{N_s}$	The ratio of the potential differences in a transformer equals the ratio of the turns

III. Know Your Units

Quantity	Dimension	SI Unit
magnetic flux (Φ)	$[M][L^2][A^{-1}][T^{-2}]$	Wb
inductance (L)	$[M][L^2][T^{-2}][A^{-2}]$	H

Puzzle

PLAY BULB!

The figure shows three identical lightbulbs attached to an ideal battery and an inductor. The switch is closed for a long time, and then opened. Which of the following statements correctly describes what happens just after the switch is opened?

1. Bulb B stays the same, and bulbs A and C go out.
2. Bulb B stays the same and bulbs A and C get brighter.
3. Bulb B goes out, and bulbs A and C get brighter.
4. Bulb B gets brighter, and bulbs A and C stay the same.
5. Bulb B gets brighter, and bulbs A and C go out.
6. All three bulbs get brighter.
7. All three bulbs go out.
8. None of the above. (State what does happen.)

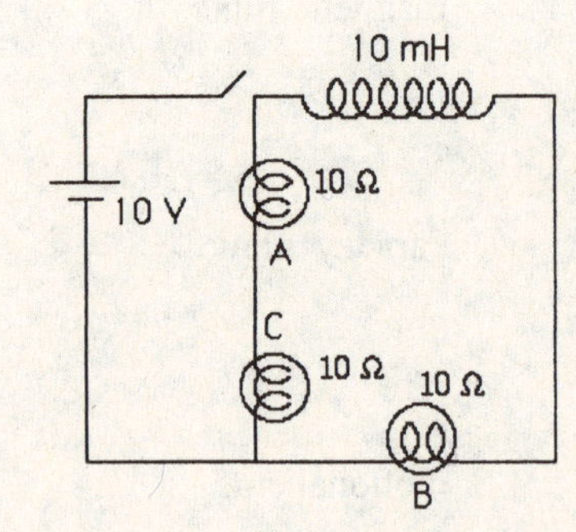

Answer this question in words, not equations, briefly explaining how you obtained your answer.

Answers to Selected Conceptual Questions and Exercises

Conceptual Questions

4. The metal plate moving between the poles of a magnet experiences eddy currents that retard its motion. This helps to damp out oscillations in the balance, resulting in more accurate readings.

10. When the angular speed of the coil in an electric generator is increased, the rate at which the magnetic flux changes increases as well. As a result, the magnitude of the induced emf produced by the generator increases. Of course, the frequency of the induced emf increases as well.

12. When the switch is opened in a circuit with an inductor, the inductor tries to maintain the original current. (In general, an inductor acts to resist *any* change in the current—whether the current is increasing or decreasing.) Therefore, the continuing current may cause a spark to jump the gap.

Conceptual Exercises

2. The induced emf is directly proportional to the rate of change of the magnetic flux, as indicated in Equation 23-4. In this system, the rate of change of magnetic flux will be proportional to the rate of change of the magnetic field. Therefore, the ranking in order of increasing magnitude of the induced emf is as follows: D = F < A < B < C < E.

8. When the disk is at its maximum displacement to the right, it is well within the region of uniform magnetic field. Therefore, the magnetic flux through the disk is not changing. It follows, then, that the induced current at this point is a minimum; namely, zero.

14. The secondary current in transformer 2 is 6 I_S. The factor of 6 is the result of a factor of 3 in the primary current, times a factor of 2 due to the doubling of the number of turns in the primary coil.

Solutions to Selected End-of-Chapter Problems

23. **Picture the Problem:** A single-resistor circuit sits next to a current-carrying wire.

Strategy: If the current in the wire changes direction, the direction of the magnetic field is reversed, changing its direction from out of the page to into the page. According to Lenz's law, the current induced in the circuit will oppose this change by flowing counterclockwise, generating a field that is directed out of the page.

31. **Picture the Problem:** A rod slides perpendicular to a magnetic field on rails that are connected to a resistor.

Strategy: The current flows clockwise, so the magnetic force is directed to the left. The external force must be equal and opposite to the magnetic force to maintain the rod's constant speed. So, it is directed to the right.

Solution:

Part (a)

1. The magnitude of the force is: $F = \dfrac{B^2 v L^2}{R}$

2. From Problem 26 the speed is: $v = \dfrac{IR}{BL}$

3. Insert the expression for v into that for F and solve: $F = \dfrac{B^2L^2}{R}\left(\dfrac{IR}{BL}\right) = IBL$

$= (0.125\ \text{A})(0.750\ \text{T})(0.45\ \text{m}) = \boxed{42\ \text{mN}}$

Part (b)

4. The rate of energy dissipation in the resistor is the power consumed by it: $P = I^2R = (0.125\ \text{A})^2(12.5\ \Omega) = \boxed{195\ \text{mW}}$

Part (c)

5. The mechanical power is delivered by the force: $P = Fv = (IBL)\left(\dfrac{IR}{BL}\right) = I^2R = \boxed{195\ \text{mW}}$

Insight Note that the conservation of energy is obeyed without any direct application of it.

37. Picture the Problem: A circular coil rotates in Earth's magnetic field

Strategy: We must consider the flux through the coil.

Solution:

Part (a) Only the $\boxed{\text{horizontal}}$ component of the magnetic field is important. The vertical component is parallel to the plane of the circular coil at all times. Thus, it does not contribute to the flux through the coil.

Part (b)

1. The maximum induced emf is: $\mathcal{E}_{\text{max}} = NBA\omega$

$$= (155)(3.80 \times 10^{-5}\text{T})\pi\left(\frac{0.220\ \text{m}}{2}\right)^2\left(1250\frac{\text{rev}}{\text{min}}\right)\left(\frac{2\pi\ \text{rad}}{1\ \text{rev}}\right)\left(\frac{1\ \text{min}}{60\ \text{s}}\right)$$

$= \boxed{29.3\ \text{mV}}$

Insight Only the magnetic field *through* the coil that contributes to the flux and therefore the induced emf

41. Picture the Problem: A solenoid has a current running through it.

Strategy: We can directly apply our results for the inductance of and induced emf in an ideal solenoid.

Solution:

Part (a)

1. From the expression for the inductance of a solenoid, we can solve for the area to get: $A = \dfrac{\ell L}{\mu_0 N^2} = \dfrac{(0.24\ \text{m})(7.3\times10^{-3}\text{H})}{\left(4\pi\times10^{-7}\ \frac{\text{T}\cdot\text{m}}{\text{A}}\right)(450)^2} = \boxed{6.9\times10^{-3}\text{m}^2}$

Part (b)

2. From Faraday's law written in terms of self-inductance: $\mathcal{E} = -L\dfrac{\Delta I}{\Delta t} = -(7.3\times10^{-3}\ \text{H})\left(\dfrac{-3.2\ \text{A}}{55\times10^{-3}\text{s}}\right) = \boxed{0.42\ \text{V}}$

Insight Compare the area found in part (a) with the size of a common object.

61. Picture the Problem: A transformer has current running through it.

Strategy: We can directly apply our results for the primary and secondary sides of a transformer.

Solution:

1. The transformer equation, solved for the primary voltage, gives: $V_p = V_s\left(\dfrac{N_p}{N_s}\right) = (4800\text{ V})\left(\dfrac{25}{750}\right) = \boxed{160\text{ V}}$

2. In terms of the relationship between the current ratio and turns ratio we have: $I_p = I_s\left(\dfrac{N_s}{N_p}\right) = (12\times10^{-3}\text{A})\left(\dfrac{750}{25}\right) = \boxed{0.36\text{ A}}$

Insight If this is a step-up transformer, why is the secondary current smaller than the primary current?

Answers to Practice Quiz

1. (a) **2.** (c) **3.** (c) **4.** (d) **5.** (d) **6.** (c) **7.** (a) **8.** (e) **9.** (b) **10.** (b) **11.** (a) **12.** (c)

CHAPTER 24
ALTERNATING-CURRENT CIRCUITS

Chapter Objectives

After studying this chapter, you should

1. know what alternating voltage and current means and how to determine their root-mean-square values.
2. be able to use phasor diagrams to represent alternating voltages.
3. understand the behavior of capacitors in AC circuits and how capacitive reactance affects the current.
4. know what the impedance in an AC circuit is and how it differs from resistance.
5. be able to calculate the phase difference between the total voltage in an AC circuit and the current.
6. understand the behavior of inductors in AC circuits and how inductive reactance affects the current.
7. be able to calculate the total voltage, impedance, current, and phase angle for an *RLC* circuit.
8. understand *LC* oscillations and resonance in *RLC* circuits.

Warm-Ups

1. Home Experiment: Make a pendulum using a string 1 to 1.5 meters long (40-60 inches), and a mass similar to a lemon (go ahead, use a lemon!). This will give you a pendulum with a resonant period in the range of 2 to 2.5 seconds ($T = 2\{L/g\}^{1/2}$). In terms of frequency, $f = 1/T$, so the resonant frequency is between 0.5 and 0.4 Hz.

   Now, hold the end of the string and vibrate it back and forth by 5 cm, slowly varying the frequency. When you find the right frequency, the lemon will swing back and forth with quite a large amplitude. This is the resonant period for your pendulum. Note the phase between the motion of your hand and the lemon. Is your hand ahead of the lemon pulling it forward, or behind pulling it back. Try swinging the lemon a little more slowly (longer period than at resonance). How does the amplitude change? How about the phase?

   Next, try going a little more quickly than at resonance. What happens to the amplitude and phase now? Finally, compare the experiment you just did with the behavior of an *RLC* circuit.

2. A radio has *RLC* circuits inside that oscillate at the same frequencies as the radio waves they receive (88.1-107.9 MHz for FM, 540-1180 KHz for AM). Select values of L and C that could be used to make these circuits.
3. An inductor or a capacitor in an AC circuit has a (time-varying) current flowing through it and voltage across it. However, the average power $P_{av} = 0$. Explain how this works out, recalling that power is equal to voltage times current.
4. Old radios used to tune from one frequency to another using variable capacitors. In such a device there are two rows of semicircular plates. One set of plates (all electrically connected together) are fixed to the base, and the other set (all connected together) can rotate through a half turn. At one extreme the rotating set are opposite to the fixed set, and at the other extreme the plates are fully "interleaved" with the fixed set. Assuming the whole thing must fit inside an old radio, estimate the maximum capacitance. (Hint: for 10 fixed and 10 rotating plates, you can treat this as 10 parallel-plate capacitors in parallel.)

Chapter Review

This chapter treats circuits with an alternating current (AC). These types of circuits are very important because AC circuits are used to provide the electricity that most of us use every day. Included with this discussion is the behavior of resistors, capacitors, and inductors in AC circuits.

24–1 Alternating Voltages and Currents

As discussed in a previous chapter, an AC generator supplies a voltage that switches (or alternates) in polarity. Typically, the voltage varies sinusoidally and can be represented by the expression

$$V = V_{\max} \sin \omega t$$

where ω is the angular frequency ($\omega = 2\pi f$) of the oscillating voltage, and $V_{\max}$ is the amplitude of the varying voltage. When this voltage is connected in a circuit with resistance R, the result is a sinusoidally **alternating current**, that is, a current that alternates in direction according to

$$I = I_{\max} \sin \omega t$$

where, by Ohm's law, $I_{\max} = V_{\max}/R$. The voltage across the resistor in the circuit and the current vary together, reaching their maxima simultaneously and equaling zero simultaneously. Because of this behavior, we say that the voltage across the resistor and the current in an AC circuit are **in phase**.

A convenient representation of alternating voltages is the use of **phasors**. A phasor is an arrow that rotates counterclockwise in an *x-y* coordinate system with an angular velocity ω equal to the angular frequency of the alternating voltage it represents. The length of the arrow represents, and is proportional to, the amplitude of the alternating voltage. The projection of the phasor onto the y-axis gives the instantaneous value of the voltage at any given time. The current in an AC circuit can also be represented by a phasor, and because the voltage across the resistor and current are in phase, the phasor for the current always points in the same direction as the phasor for this voltage.

Because the voltage and current in an AC circuit alternate, it is customary to characterize quantities by an average value. However, the average values of the voltage and current are zero so instead we use the **root-mean-square** (rms) values. An rms value of a quantity is the square root of the average (mean) of the squared quantity. In the case of sinusoidally varying quantities, the rms value works out to be the maximum value divided by $\sqrt{2}$:

$$V_{rms} = V_{\max} / \sqrt{2}$$
$$I_{rms} = I_{\max} / \sqrt{2}$$

Using rms values, we can write equations for AC circuit that mirror equations we used for DC circuits. For example, an AC version of Ohm's law for a purely resistive circuit is

$$V_{rms} = I_{rms}R$$

Also, the average power consumed by the resistance R in an AC circuit is conveniently written, in terms of rms values, by expressions that look just like the ones we used with DC circuits:

$$P_{av} = I_{rms}^2 R = I_{rms}V_{rms} = V_{rms}^2 / R$$

The instantaneous power used in the resistor is determined by the same expressions, except we replace I_{rms} with I, and V_{rms} with V. Also, any expression involving rms values can also be written for maximum values ($V_{\max}$, $I_{\max}$, $P_{\max}$), simply by replacing the rms value of the quantity with its corresponding maximum value.

Example 24–1 AC A simple circuit contains an AC generator with a maximum output of 150 V operating at a frequency of 60 Hz. If the resistance in the circuit is 35 Ω, what average power is dissipated through the resistor?

Picture the Problem The diagram shows an AC generator connected to a resistor.

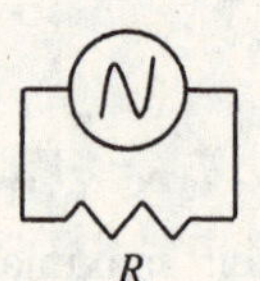

Strategy Of the three equivalent expressions for P_{av} we choose the one most convenient for the given information. Since we are given R and $V_{\max}$, we will use $P_{av} = V_{rms}^2 / R$.

Solution

1. The rms voltage is: $V_{rms} = V_{max}/\sqrt{2} = (150\text{ V})/\sqrt{2} = 106.1\text{ V}$

2. The average power then is: $P_{av} = V_{rms}^2/R = (106.1\text{ V})^2/35\,\Omega = 320\text{ W}$

Insight Any of the three expressions could have been used to determine P_{av}, but this one was most convenient because it involved the fewest intermediate steps.

Practice Quiz

1. An alternating current is given by $I = (0.25\text{ A})\sin[(377\text{ rad/s})\,t]$. What is the rms current?
   (a) 0.25 A **(b)** 0.18 A **(c)** 377 A **(d)** 0.12 A **(e)** 0.50 A

2. An alternating current is given by $I = (0.25\text{ A})\sin[(377\text{ rad/s})\,t]$. What is the frequency f?
   (a) 377 Hz **(b)** 0.25 Hz **(c)** 60.0 Hz **(d)** 0.18 Hz **(e)** 2370 Hz

23–2 – 23–3 Capacitors in AC Circuits and RC Circuits

If you consider a circuit consisting of just an AC generator and a capacitor, with no resistance in it, the capacitor itself, because of charging and discharging, offers some opposition to the current being sent by the generator. This resistance-like opposition to the current due to the charging and discharging of the capacitor in an AC circuit is called the **capacitive reactance,** X_C; it is given by

$$X_C = \frac{1}{\omega C}$$

The capacitive reactance has the same SI unit as resistance, the ohm, Ω. Written in terms of the capacitive reactance, the rms current in the circuit is

$$I_{rms} = \frac{V_{rms}}{X_C}$$

Recall that in a DC circuit the capacitor takes time to charge and discharge; similar behavior occurs in AC circuits, as well. Because of this fact, there is a phase difference between the current and the voltage across the capacitor:

the voltage across a capacitor lags the current by 90°.

Because of this lag, the phasor for the voltage across the capacitor is drawn 90° behind the phasor for the current, with both phasors rotating counterclockwise, as shown below.

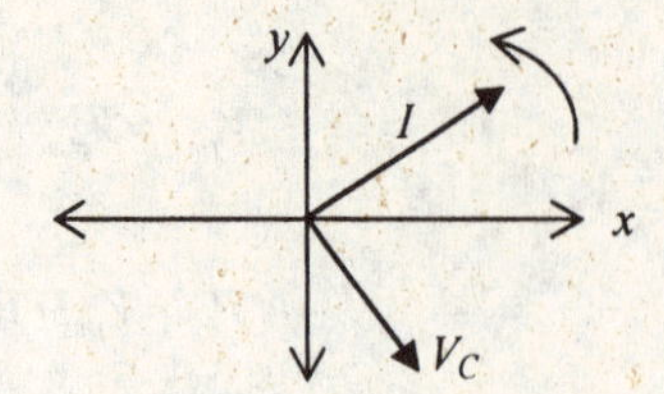

The key differences between the capacitive reactance and resistance are that (a) they result from very different types of physical processes; (b) the capacitive reactance depends on the frequency of the generator, getting smaller as ω increases; and (c) there is no net power consumption associated with X_C, as there is with R, because no net energy is used by the capacitor.

If we now consider an *RC* circuit run by an AC generator, we must combine the considerations of Section 24–1 with the present information. The voltage across the resistor, V_R, is in phase with the current and is 90° out of phase with the voltage across the capacitor. The phasors for these two voltages can be used to form a right triangle, and the total voltage in the circuit can be determined using the Pythagorean theorem. For maximum values we have

$$V_{\max} = \sqrt{V_{\max,R}^2 + V_{\max,C}^2}$$

This expression can also be written as $V_{\max} = I_{\max}\sqrt{R^2 + X_C^2}$. Making the definition

$$Z = \sqrt{R^2 + X_C^2}$$

we can write this relation in a way that look's similar to Ohm's law:

$$V_{\max} = I_{\max} Z$$

The quantity Z is called the **impedance** of the circuit and represents the combined resistance-like effects of the actual resistance and the capacitance. The SI unit of impedance is the ohm. The preceding expressions can also be written for rms values.

In general, the total voltage in an RC circuit will be out of phase with the current by an angle ϕ between 0 and –90°, where the minus indicates that the voltage lags the current. Using trigonometry on the phasor diagram gives

$$\tan\phi = \frac{X_C}{R} \quad \text{and} \quad \cos\phi = \frac{R}{Z}$$

these expressions can be used to find the magnitude of the angle ϕ. Because energy is only consumed by the resistor, the phase angle also comes into play in determining the average power consumed in the circuit. The result, equivalent to $P_{av} = I_{rms}^2 R$, is

$$P_{av} = I_{rms} V_{rms} \cos\phi$$

The factor cos ϕ is called the **power factor**.

Example 24–2 RC Circuits An AC circuit contains a 44.3-μF capacitor in series with a 50.1-Ω resistor. The circuit is powered by a 60.0-Hz generator with an rms output of 85.0 V. Find the rms current in the circuit and the average power consumed through the resistor.

Picture the Problem The diagram shows a series *RC* circuit with an AC generator.

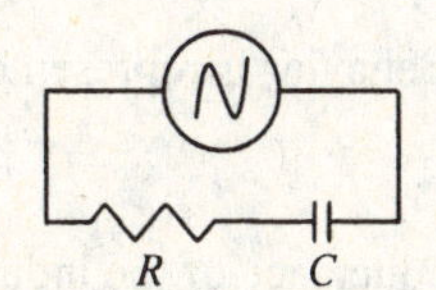

Strategy We start with the basic expression for I_{rms} and find all the necessary quantities for that expression. Then, we should have enough information to determine P_{av}.

Solution

1. An expression for I_{rms} is: $$I_{rms} = \frac{V_{rms}}{Z} = \frac{V_{rms}}{\sqrt{R^2 + X_C^2}}$$

2. The capacitive reactance is: $$X_C = \frac{1}{\omega C} = \frac{1}{2\pi fC} = \frac{1}{2\pi(60.0\text{ Hz})(44.3\ \mu\text{F})} = 59.88\ \Omega$$

3. The current then is: $$I_{rms} = \frac{V_{rms}}{\sqrt{R^2 + X_C^2}} = \frac{85.0\text{ V}}{\left[(50.1\ \Omega)^2 + (59.88\ \Omega)^2\right]^{\frac{1}{2}}} = 1.09\text{ A}$$

4. The average power can be found as: $$P_{av} = I_{rms}^2 R = (1.089\text{ A})^2(50.1\ \Omega) = 59.4\text{ W}$$

Insight The formula used for the rms current is just a rearrangement of the expression involving the maximum current and voltage, with the maximum values replaced by rms values.

Practice Quiz

3. An *RC* circuit contains a 25-Ω resistor and a 56-μF capacitor. The circuit is run by a 60-Hz AC generator with a maximum emf of 120 V. What is the impedance of this circuit?

   (a) 47 Ω **(b)** 25 Ω **(c)** 380 Ω **(d)** 54 Ω **(e)** 60 Ω

4. An RC circuit contains a 25-Ω resistor and a 56-μF capacitor. The circuit is run by a 60-Hz AC generator with a maximum emf of 120 V. What is the rms current in this circuit?

(a) 1.6 A **(b)** 2.2 A **(c)** 3.4 A **(d)** 1.8 A **(e)** 2.5 A

24–4 Inductors in AC Circuits

For an inductor in an AC circuit, without resistance, the induced emf delays the current in reaching its maximum, which means that the inductor offers a resistance-like opposition to the current sent by the generator. This behavior is represented by what is called the **inductive reactance**, which is given by

$$X_L = \omega L$$

where L is the inductance of the inductor. With this result, the rms current in the circuit becomes

$$I_{rms} = \frac{V_{rms}}{X_L}$$

Recall that the induced emf across the inductor is greatest when the current is changing most rapidly. This maximum rate occurs at the zero point of the current oscillation. Thus, the voltage across the inductor is out of phase with the current. In fact,

the voltage across the inductor V_L leads the current by 90°.

Because of this lead, the phasor for the voltage across the inductor is drawn 90° ahead of the phasor for the current, with both phasors rotating counterclockwise, as shown below.

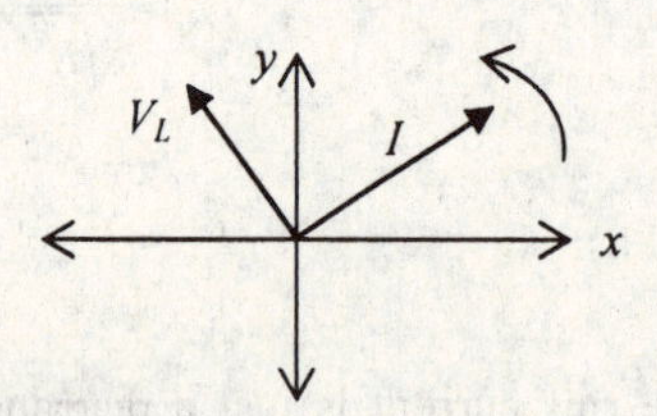

The key differences between the inductive reactance and resistance are that (a) they result from very different types of physical processes; (b) the inductive reactance depends on the frequency of the generator, getting larger as ω increases; and (c) there is no net power consumption associated with X_L, as there is with R, because no net energy is used by an ideal inductor.

If we now consider an RL circuit run by an AC generator, we notice that the voltage across the resistor lags the voltage across the inductor by 90°. The phasors for these two voltages, therefore, can be used to form a right triangle, and the total voltage in the circuit is determined using the Pythagorean theorem. The result is

$$V_{\max} = \sqrt{V_{\max,R}^2 + V_{\max,L}^2}$$

This expression can also be written as $V_{\max} = I_{\max}\sqrt{R^2 + X_L^2}$. Here, the impedance is given by

$$Z = \sqrt{R^2 + X_L^2}$$

and we can write this relation as

$$V_{\max} = I_{\max} Z$$

As with the *RC* circuit, in general, the total voltage in an *RL* circuit will be out of phase with the current. However, for the case of an *RL* circuit, the phase angle ϕ is between 0 and +90° because the voltage leads the current in this case. The phase angle can be determined from

$$\tan\phi = \frac{X_L}{R} \quad \text{and} \quad \cos\phi = \frac{R}{Z}$$

Exercise 24–3 An *RL* Circuit An AC circuit contains a 40-mH inductor in series with a 35-Ω resistor. The circuit is powered by a 25-Hz generator with a maximum output of 115 V. Find the rms voltage across the inductor.

Solution: We are given the following information:

Given: $L = 40$ mH, $R = 35\ \Omega$, $f = 25$ Hz, $V_{\max} = 115$ V; **Find:** $V_{rms,L}$

The relationship for the rms voltage across the inductor is

$$V_{rms,L} = I_{rms} X_L$$

To calculate this voltage we need to find both I_{rms} and X_L. Because we know the frequency and the inductance, the inductive reactance is

$$X_L = \omega L = 2\pi f L = 2\pi(25\text{ Hz})(40\times10^{-3}\text{ H}) = 6.28\ \Omega$$

The expression for the rms current is

$$I_{rms} = \frac{V_{rms}}{Z} = \frac{V_{rms}}{\sqrt{R^2 + X_L^2}}$$

Using the given information we get

$$I_{rms} = \frac{V_{\max}/\sqrt{2}}{\sqrt{R^2 + X_L^2}} = \frac{(115\,V)/\sqrt{2}}{\sqrt{(35\ \Omega)^2 + (6.28\ \Omega)^2}} = 2.287\text{ A}$$

Now, we are able to calculate the rms voltage across the inductor to be

$$V_{rms,L} = (2.287\text{ A})(6.28\ \Omega) = 14\text{ V}$$

This result is only the rms voltage across the inductor. It is considerably smaller than the total rms voltage because of the relatively low frequency of the generator.

Practice Quiz

5. An RL circuit contains a 25-Ω resistor and a 56-mH inductor. The circuit is run by a 60-Hz AC generator with a maximum emf of 120 V. What is the inductive reactance in this circuit?

(a) 60 Ω **(b)** 0.35 Ω **(c)** 377 Ω **(d)** 25 Ω **(e)** 21 Ω

6. An RL circuit contains a 25-Ω resistor and a 56-mH inductor. The circuit is run by a 60-Hz AC generator with a maximum emf of 120 V. What is the maximum current in this circuit?

(a) 4.8 A **(b)** 3.7 A **(c)** 2.6 A **(d)** 5.7 A **(e)** 3.4 A

24–5 *RLC* Circuits

In this section we combine all the above results to consider an AC circuit containing a resistor, an inductor, and a capacitor in series — an ***RLC* circuit**. The phasors for the voltage in this circuit show that V_L leads the V_R by 90°, and V_C lags V_R by 90°. Therefore, V_C lags V_L by 180°, and we can consider their combined effect as $V_L - V_C$. The total maximum voltage is given by

$$V_{\max} = \sqrt{V_{\max,R}^2 + \left(V_{\max,L} - V_{\max,C}\right)^2}$$
$$= I_{\max}\sqrt{R^2 + \left(X_L - X_C\right)^2}$$

This shows that for an RLC circuit the impedance is

$$Z = \sqrt{R^2 + \left(X_L - X_C\right)^2}$$

The phase angle between the total voltage in the circuit and the current is determined by

$$\tan\phi = \frac{X_L - X_C}{R} \quad \text{and} \quad \cos\phi = \frac{R}{Z}$$

It is worth remembering that the power factor cos ϕ will only give the magnitude of ϕ; it does not tell you if the voltage leads or lags the current. Expressing the phase angle in terms of the tangent, however, gives both the magnitude and sign of ϕ. If ϕ is positive, the total voltage leads the current, if it is negative the voltage lags the current.

Example 24–4 An RLC Circuit An *RLC* circuit has $R = 55.5\ \Omega$, $L = 15.6$ mH, and $C = 75.2\ \mu$F. If the AC generator connected to this circuit has rms output of 120.0 V and operates at a frequency of 60.0 Hz, find **(a)** the impedance of the circuit, **(b)** the rms current in the circuit, **(c)** the rms voltage across the resistor, **(d)** the rms voltage across the inductor, **(e)** the rms voltage across the capacitor, **(f)** the phase angle between the total voltage and the current, and **(g)** the power consumed through the resistor.

Picture the Problem The diagram shows an *RLC* circuit connected to an AC generator.

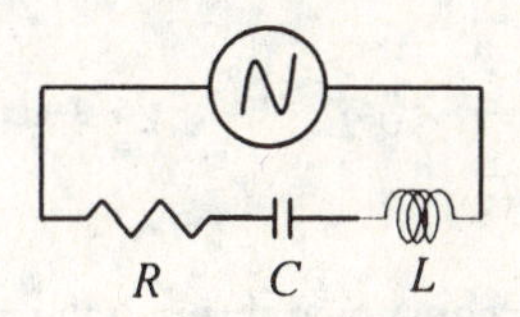

Strategy For each part, we use the expression for the desired quantity as a guide for what to calculate in intermediate steps.

Solution

Part (a)

1. The expression for the impedance shows that we first need X_L and X_C:

$$Z = \sqrt{R^2 + (X_L - X_C)^2}$$

2. The inductive reactance is:

$$X_L = \omega L = 2\pi fL$$
$$= 2\pi(60.0\text{ Hz})(15.6\times10^{-3}\text{ H}) = 5.881\ \Omega$$

3. The capacitive reactance is:

$$X_C = \frac{1}{\omega C} = [2\pi fC]^{-1}$$
$$= \left[2\pi(60.0\text{ Hz})(75.2\times10^{-6}\text{ F})\right]^{-1} = 35.27\ \Omega$$

4. The impedance then becomes:

$$Z = \sqrt{(55.5\ \Omega)^2 + (5.881\ \Omega - 35.27\ \Omega)^2} = 62.8\ \Omega$$

Part (b)

5. The rms current is:

$$I_{rms} = \frac{V_{rms}}{Z} = \frac{120\text{ V}}{62.80\ \Omega} = 1.91\text{ A}$$

Part (c)

6. The rms voltage across the resistor is:

$$V_{rms,R} = I_{rms}R = (1.911\text{ A})(55.5\ \Omega) = 106\text{ V}$$

Part (d)

7. The rms voltage across the inductor is:

$$V_{rms,L} = I_{rms}X_L = (1.911\text{ A})(5.881\ \Omega) = 11.2\text{ V}$$

Part (e)

8. The rms voltage across the capacitor is: $V_{rms,C} = I_{rms}X_C = (1.911\ \text{A})(35.27\ \Omega) = 67.4\ \text{V}$

Part (f)

9. The phase angle is given by: $\tan\phi = \dfrac{X_L - X_C}{R} \Rightarrow \phi = \tan^{-1}\left(\dfrac{X_L - X_C}{R}\right)$

$$\therefore\ \phi = \tan^{-1}\left(\frac{5.881\ \Omega - 35.27\ \Omega}{55.5\ \Omega}\right) = -27.9^\circ$$

Part (g)

10. The power consumed through the resistor is: $P_{av} = I_{rms}^2 R = (1.911\ \text{A})^2(55.5\ \Omega) = 203\ \text{W}$

Insight Once all the reactances and the impedance are known, the rest of the results are obtained rather easily. Notice that the sum of the rms voltages across the resistor, inductor, and capacitor is greater than the total rms voltage of 120 V. This fact is not a problem because these voltages are out of phase, so they reach their maxima (or rms) values at different times. The negative phase angle means that the total voltage lags the current. Can you sketch the phasor diagram?

Practice Quiz

7. An *RLC* circuit contains a 25-Ω resistor, a 56-mH inductor, and an 85-μF capacitor; it is run by a 60-Hz AC generator with a maximum emf of 120 V. What is the impedance of this circuit?

   (a) 25 Ω **(b)** 21 Ω **(c)** 31 Ω **(d)** 36 Ω **(e)** 27 Ω

8. An *RLC* circuit contains a 25-Ω resistor, a 56-mH inductor, and an 85-μF capacitor; it is run by a 60-Hz AC generator with a maximum emf of 120 V. What is the phase angle between the current and the total voltage?

   (a) +22° **(b)** +47° **(c)** −22° **(d)** −47° **(e)** 0.0°

24–6 Resonance in Electrical Circuits

In a pure *LC* circuit, with an initially charged capacitor, no resistor, and no generator, the current will still oscillate between charging the capacitor and flowing through the inductor. This also means that the energy contained in the circuit oscillates between the electric field of the capacitor and the magnetic field of the inductor. These oscillations naturally take place at a characteristic frequency that depends on the values of the inductance and capacitance. The natural frequency of oscillation for an ideal *LC* circuit is given by

$$\omega = \frac{1}{\sqrt{LC}}$$

The fact that current oscillates with a natural frequency in an *LC* circuit leads to the phenomenon of resonance in *RLC* circuits. The AC generator is what drives the oscillation in an *RLC* circuit. When the generator drives the circuit at the natural frequency the current has its largest amplitude, and we have achieved resonance in this circuit. The current will be largest when the impedance is smallest. The smallest impedance is $Z = R$, which occurs when the frequency in the circuit is such that $X_L = X_C$. This occurs at precisely the natural frequency for the *LC* oscillations. These results also mean that the phase angle between the current and the total voltage is zero; that is, the current and voltage are in phase at resonance.

Example 24–5 Resonance An *RLC* circuit has $R = 55.5\ \Omega$, $L = 15.6$ mH, and $C = 75.2\ \mu$F. If the AC generator of this circuit has an rms output of 120.0 V and operates at a frequency of 60.0 Hz, **(a)** what is the resonant frequency of this circuit, and **(b)** what would be the average power consumed through the resistor if the generator was operated at the resonant frequency?

Picture the Problem The diagram shows an *RLC* circuit connected to an AC generator.

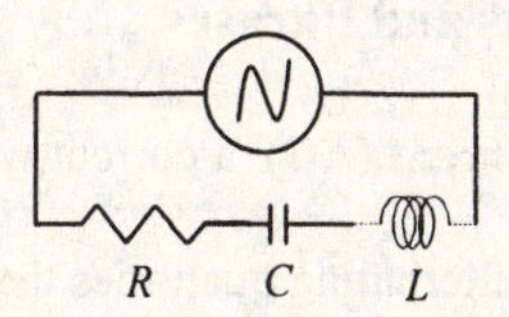

Strategy The resonant frequency is the natural frequency for the *LC* oscillation. We can use our knowledge of what happens at resonance to simplify the calculation of the average power.

Solution

Part (a)

1. The resonant frequency is (in rad/s): $\omega = 1/\sqrt{LC} = \left[(15.6\times10^{-3}\ \text{H})(75.2\times10^{-6}\ \text{F})\right]^{-1/2} = 923\ \text{rad/s}$

Part (b)

2. At resonance $Z = R$, therefore: $V_{rms} = I_{rms}Z = I_{rms}R = V_{rms,R}$

3. The average power is therefore: $P_{av} = \dfrac{V_{rms,R}^2}{R} = \dfrac{V_{rms}^2}{R} = \dfrac{(120.0\ \text{V})^2}{55.5\ \Omega} = 259\ \text{W}$

Insight There are other routes to determining the average power at resonance. Think of at least one other way, and then try it.

Practice Quiz

9. An *RLC* circuit contains a 25-Ω resistor, a 56-mH inductor, and an 85-μF capacitor; it is run by a 60-Hz AC generator with a maximum emf of 120 V. What is the resonant frequency (in Hertz) of this circuit?

   (a) 73 Hz (b) 460 Hz (c) 2.2 Hz (d) 60 Hz (e) 380 Hz

10. An *RLC* circuit contains a 15-Ω resistor, a 20-mH inductor, and a 75-μF capacitor; it is run by an AC generator of variable frequency. At what frequency (either ω or f) will the power output of this circuit be maximum?

    (a) 60 Hz (b) 380 rad/s (c) 820 Hz (d) 130 Hz (e) 30 rad/s

Reference Tools and Resources

I. Key Terms and Phrases

alternating current (AC) a current whose direction alternates in a circuit

in phase two alternating quantities that reach their maxima and minima simultaneously oscillate in phase

phasor an arrow in an *x-y* coordinate system that rotates counterclockwise to represent alternating voltage or current

root-mean-square (rms) the square root of the mean of a squared quantity

capacitive reactance the resistance-like behavior of a capacitor in an AC circuit

inductive reactance the resistance-like behavior of an inductor in an AC circuit

impedance the resistance-like behavior of an AC circuit that combines the effects of the resistance, capacitive reactance and inductive reactance

RLC circuit an AC circuit containing a resistor, a capacitor, and an inductor in series

power factor a multiplicative factor, cos ϕ, that determines the power consumed through the resistor in an *RLC* circuit

II. Important Equations

Name/Topic	Equation	Explanation
alternating voltages and currents	$V = V_{max}\sin\omega t$ $I = I_{max}\sin\omega t$	Equations representing sinusoidally varying voltage and current
capacitive reactance	$X_C = \frac{1}{\omega C}$	The definition of the capacitive reactance
inductive reactance	$X_L = \omega L$	The definition of the inductive reactance
impedance	$Z = \sqrt{R^2 + (X_L - X_C)^2}$	The impedance of an *RLC* circuit
phase angle	$\tan\phi = \frac{X_L - X_C}{R}$	The tangent of the phase angle allows us to calculate ϕ for an *RLC* circuit
AC voltage	$V_{max} = I_{max}Z$	The expression similar to Ohm's law for an AC circuit
natural frequency	$\omega = \frac{1}{\sqrt{LC}}$	The natural frequency of *LC* oscillation that produces resonance in an *RLC* circuit

III. Know Your Units

Quantity	Dimension	SI Unit
reactance and impedance (X_C, X_L, Z)	$[M][L^2][A^{-2}][T^{-3}]$	Ω

IV. Tips

There is an old mnemonic that people sometimes use to remember the phase relationships between the current and the voltages across inductors and capacitors in AC circuits. The mnemonic is based on the phrase

ELI the **ICE** man

and the associations **E** for emf (or voltage), **I** for current, **L** for inductor, and **C** for capacitor.

The word **ELI** contains the letter **L** and, therefore, refers to the inductor. Notice that the **E** in **ELI** comes before the **I**, which serves to remind you that the voltage across the inductor "comes before" (or leads) the current. The word **ICE** contains the letter **C** and, therefore, refers to the capacitor. Since the **E** in **ICE** comes after the **I**, this word reminds you that the voltage across the capacitor "comes after" (or lags) the current. If you find mnemonic devices such as this useful, then this is a good one to use.

Puzzle

UPS AND DOWNS

The figure shows a graph of the current in a series *RLC* circuit, $I = I_{max}\cos(t)$. There are five points marked on the graph, labeled A, B, C, D, E. State at which point (or points) each of the following quantities is a maximum, a minimum, or zero:

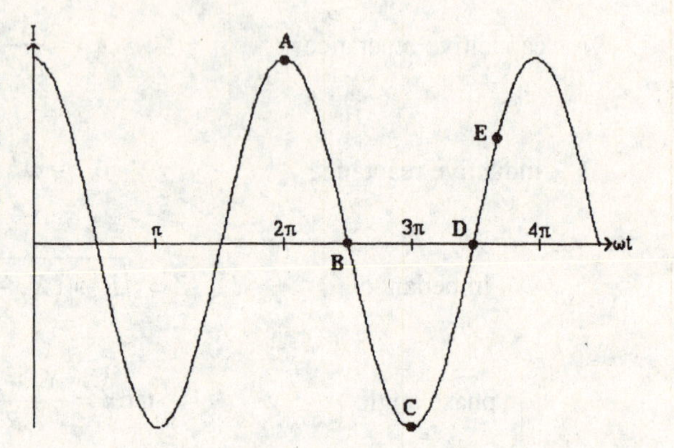

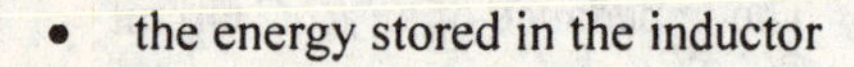

- the energy stored in the inductor
- the voltage across the capacitor
- the power input to the capacitor
- the power output by the inductor
- the charge on the capacitor
- the voltage across the resistor

Answers to Selected Conceptual Questions and Exercises

Conceptual Questions

2. Current and voltage are not always in phase in an ac circuit because capacitors and inductors respond not to the current itself – as a resistor does – but to the charge (capacitor) or to the rate of change of the current (inductor). The charge takes time to build up; therefore, a capacitor's voltage lags behind the current. The rate of change of current is greatest when the current is least; therefore, an inductor's voltage leads the current. Resistors, of course, are always in phase with the current.

8. Recall that charge in an *RLC* circuit is the analog of position in a mass-spring system. Therefore, the current – which is the rate of change of charge – is analogous to the velocity – which is the rate of change of position. (See Table 24-2.)

10. Yes. All that is required for their resonance frequencies to be the same is for the product of *L* and *C* to be the same. (See Equation 24-18.)

Conceptual Exercises

2. Light will attain its maximum brightness 120 times per second; that is, twice per cycle. The reason for the factor of two per cycle is that the current reverses direction every cycle, and the bulb will be brightest when the current is a maximum in either direction.

6. At low frequency, the capacitor is essentially the same as a break in the circuit, whereas the inductor is essentially an ideal wire. It follows, then, that more current will be supplied by the generator if the inductor and the capacitor are connected in parallel.

10. At high frequency, we can replace the inductor with an open circuit. At low frequency, we can replace the capacitor with an open circuit. In either case, the effective resistance of the circuit is R; therefore, the current is the same at both high and low frequency.

Solutions to Selected End-of-Chapter Problems

5. Picture the Problem: As the voltage across a resistor oscillates, power is dissipated in the resistor at a rate proportional to the square of the voltage.

Strategy: Solve Eq. 24-6 for the average power dissipation and solve Eq. 21-6 for the peak power dissipation.

Solution: 1. (a) Calculate the average power dissipation: $P_{av} = \frac{1}{2}\frac{V_{max}^{2}}{R} = \frac{1}{2}\frac{(121\text{ V})^2}{(3.33\times10^3\,\Omega)} = \boxed{2.20\text{ W}}$

2. (b) Calculate the peak power dissipation: $P_{max} = \frac{V_{max}^{2}}{R} = \frac{(121\text{ V})^2}{3.33\times10^3\,\Omega} = \boxed{4.40\text{ W}}$

Insight: The peak power dissipation in a resistor is double the average power dissipation for a sinusoid voltage across the resistor.

13. Picture the Problem: The image shows a 22 μF capacitor connected to a voltage source oscillating at 120 Hz. The maximum current in the circuit is 0.15 A.

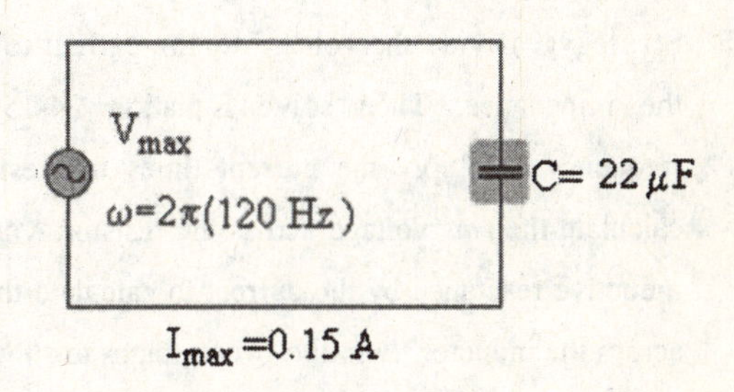

Strategy: Calculate the capacitive reactance using Eq. 24-9 and insert it into Eq. 24-8 to calculate the maximum voltage. Calculate the phase angle for which the current is 0.10 A using Eq. 24-2. Insert the phase angle into the equation $V = V_{max}\sin(\theta - 90°)$, to calculate the voltage.

Solution: 1. (a) Calculate the capacitive reactance: $X_C = \frac{1}{\omega C} = \frac{1}{\left[2\pi\left(120\text{ s}^{-1}\right)\right]22\,\mu\text{F}} = 60.29\,\Omega$

2. Insert the reactance into Equation 24-8: $V_{max} = X_C I_{max} = 60.29\,\Omega(0.15\text{ A}) = \boxed{9.0\text{ V}}$

3. (b) Calculate the phase:

$I = I_{max} \sin\theta$

$\theta = \sin^{-1}\dfrac{I}{I_{max}} = \sin^{-1}\dfrac{0.10\text{ A}}{0.15\text{ A}} = 41.8° \text{ or } 138°$

4. Calculate the voltage when the current is rising ($\theta = 41.8°$):

$V = V_{max}\sin(\theta - 90°) = 9.04\text{ V}\sin(41.8° - 90°) = \boxed{-6.7\text{ V}}$

5. (c) Calculate the voltage when the current is decreasing ($\theta = 138°$):

$V = 9.04\text{ V}\sin(138° - 90°) = \boxed{6.7\text{ V}}$

Insight: As the current is increasing, the voltage is increasing from its maximum negative value to zero. As the voltage increases from zero to the maximum positive value, the current is positive, but decreasing.

25. Picture the Problem: The figure shows a 105 Ω resistor and 82.4 μF capacitor connected in series with a 125 V, 70.0 Hz power supply.

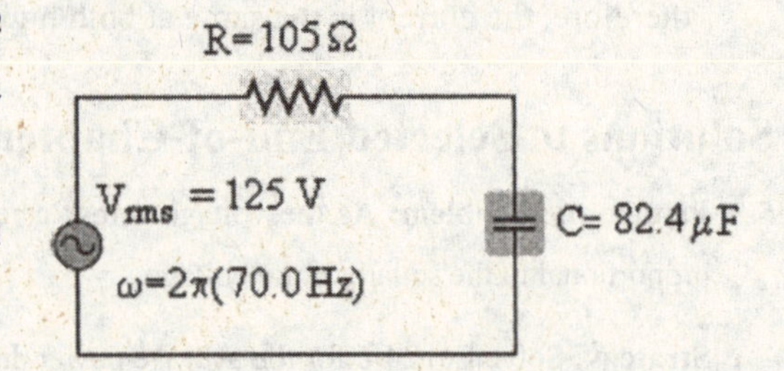

Strategy: Calculate the impedance using Eq. 24-11 and insert this into Eq. 24-12 to calculate the power factor.

Solution: 1. Calculate the impedance: $Z = \sqrt{R^2 + (\omega C)^{-2}} = \sqrt{(105\ \Omega)^2 + \left[2\pi(70\text{ Hz})(82.4\mu F)\right]^{-2}} = 108.6\ \Omega$

2. Calculate the power factor: $\cos\phi = \dfrac{R}{Z} = \dfrac{105\ \Omega}{108.6\ \Omega} = \boxed{0.967}$

Insight: The *rms* current in this circuit is 1.15 A and the average power dissipation is 140 W.

37. Picture the Problem: The image shows a 68 Ω resistor and a 31 mH inductor connected in series with a 120 V oscillating power source.

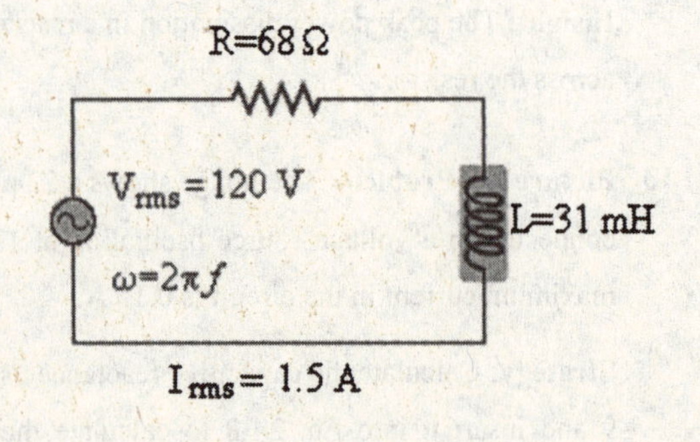

Strategy: Divide the voltage by the current to calculate the impedance. Then solve Equation 24-15 for the frequency. Multiply the current times the resistance to calculate the *rms* voltage across the resistor. Multiply the inductive reactance by the current to calculate the voltage across the inductor. Sum the two voltages to show that the sum is greater than 120 V. Sum the squares of the voltages and take the square root to show that this is equal to the voltage from the power supply.

Solution: 1. (a) Divide the voltage by the current:

$Z = \dfrac{V_{rms}}{I_{rms}} = \dfrac{120\text{ V}}{1.5\text{ A}} = 80\ \Omega,$

2. Solve Equation 24-15 for the angular frequency:

$$Z = \sqrt{R^2 + (\omega L)^2}$$

$$\omega = \frac{\sqrt{Z^2 - R^2}}{L} = \frac{\sqrt{(80\ \Omega)^2 - (68\ \Omega)^2}}{31\text{ mH}} = 1360\text{ rad/s}$$

3. Divide the angular frequency by 2π:

$$f = \frac{\omega}{2\pi} = \frac{1359\text{ rad/s}}{2\pi} = \boxed{0.22\text{ kHz}}$$

4. **(b)** Multiply the current times resistance:

$$V_{\text{rms},R} = I_{\text{rms}}R = 1.5\text{ A}(68\ \Omega) = \boxed{0.10\text{ kV}}$$

5. **(c)** Multiply the current times reactance:

$$V_{\text{rms},L} = I_{\text{rms}}X_{\text{L}} = I_{\text{rms}}\omega L = 1.5\text{ A}(1359\text{ rad/s})(31\text{ mH}) = \boxed{63\text{ V}}$$

6. **(d)** Sum the *rms* voltages:

$$V_{\text{rms},R} + V_{\text{rms},L} = 102\text{ V} + 63\text{ V} = 165\text{ V} > 120\text{ V}$$

7. Calculate the square root of the sum of the squares of the voltages:

$$\sqrt{V_{\text{rms},R}^2 + V_{\text{rms},L}^2} = \sqrt{(102\text{ V})^2 + (63\text{ V})^2} = 120\text{ V}$$

Insight: Because the voltage across the inductor and resistor are out of phase their average voltages sum greater than the source voltage. The instantaneous voltage across the inductor and resistor still sums to the instantaneous voltage across the power source.

47. Picture the Problem: The figure shows a 105 Ω resistor, 85.0 mH inductor, and 13.2μF capacitor connected in series with a 125 Hz power supply.

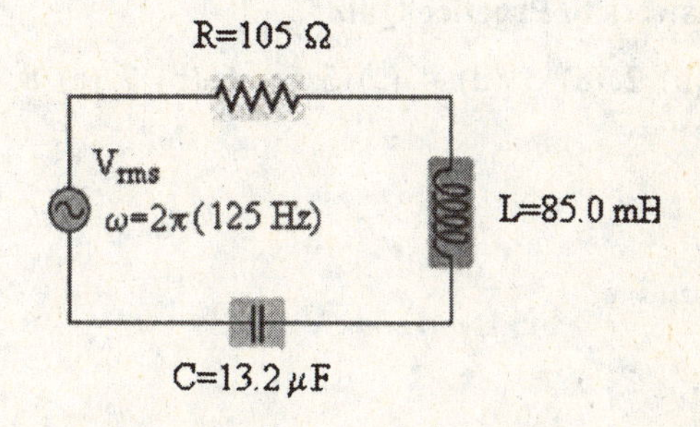

Strategy: Calculate the capacitive reactance (Eq. 24-9) and the inductive reactance (Eq. 24-14). Use these in Eq. 24-17 to calculate the phase angle. Take the cosine of the phase angle to calculate the power factor.

Solution: 1: Calculate the capacitive reactance:

$$X_{\text{C}} = \frac{1}{\omega C} = \frac{1}{2\pi(125\text{ Hz})13.2\ \mu\text{F}} = 96.46\ \Omega$$

2. Calculate the inductive reactance:

$$X_{\text{L}} = \omega L = 2\pi(125\text{ Hz})(85\text{ mH}) = 66.76\ \Omega$$

3. Solve Equation 24-17 for the phase angle:

$$\phi = \tan^{-1}\frac{X_L - X_C}{R} = \tan^{-1}\frac{66.76\ \Omega - 96.46\ \Omega}{105\ \Omega} = -15.79°$$

4. Calculate the power factor:

$$\cos\phi = \cos(-15.79°) = \boxed{0.962}$$

5. **(b)** As *R* increases, magnitude of the phase angle decreases. As the magnitude of the phase angle decreases the cosine increases, and thus, the power factor $\boxed{\text{increases}}$.

6. **(c)** Calculate the power factor when *R*=525 Ω:

$$\cos\phi = \cos\left[\tan^{-1}\frac{X_L - X_C}{R}\right] = \cos\left[\tan^{-1}\frac{66.76\ \Omega - 96.46\ \Omega}{525\ \Omega}\right] = \boxed{0.998}$$

Insight: Decreasing the magnitude of the resistance will decrease the power factor. A resistance of 30Ω gives a power factor of 0.71

55. Picture the Problem: The figure shows a 115 Ω resistor, 57.6 mH inductor, and a 179 μF capacitor connected in series with an oscillating power source.

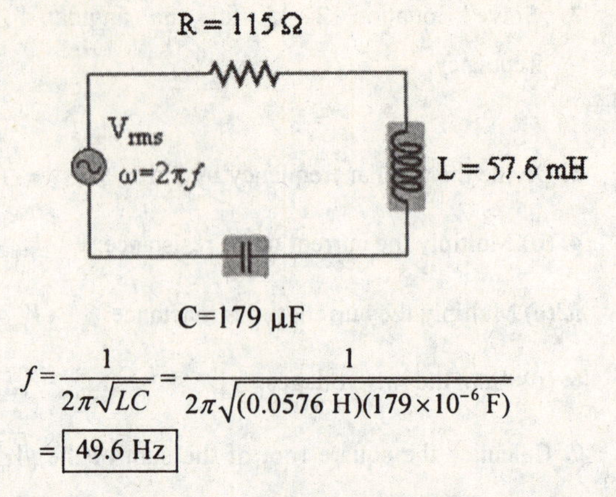

Strategy: The current will be at its maximum and the impedance will be at its minimum when the frequency is at the resonance frequency. Solve Equation 24-18 for the resonance frequency.

Solution: (a) & (b): Calculate the resonance frequency:

$$f = \frac{1}{2\pi\sqrt{LC}} = \frac{1}{2\pi\sqrt{(0.0576 \text{ H})(179\times10^{-6} \text{ F})}} = \boxed{49.6 \text{ Hz}}$$

Insight: At resonant frequency the inductive reactance and the capacitive reactance are equal. As such, the impedance is a minimum and equal to the resistance (115 Ω) and the current is equal to the voltage divided by the resistance.

Answers to Practice Quiz

1. (b) **2.** (c) **3.** (d) **4.** (a) **5.** (e) **6.** (b) **7.** (e) **8.** (c) **9.** (a) **10.** (d)

CHAPTER 25

ELECTROMAGNETIC WAVES

Chapter Objectives

After studying this chapter, you should

1. know what makes up an electromagnetic wave.
2. know the relationship between the directions of $\vec{\mathbf{E}}$, $\vec{\mathbf{B}}$, and the direction of propagation of the wave.
3. be able to calculate the Doppler shift for electromagnetic waves.
4. be familiar with the regions of the electromagnetic spectrum.
5. know the relationship among the wavelength, frequency, and speed of an electromagnetic wave.
6. be able to calculate the energy density, intensity, and radiation pressure of an electromagnetic wave.
7. know the relationship between the magnitudes of $\vec{\mathbf{E}}$ and $\vec{\mathbf{B}}$, and the speed of light.
8. be able to use the law of Malus to determine the intensity of light passing through a polarizer.

Warm-Ups

1. In a wave on a string, the amplitude is the maximum displacement of a point on the string from its equilibrium position. What is the amplitude of an electromagnetic wave?
2. Estimate the wavelength of your favorite radio station.
3. At a given distance, say 3 m, the light from a 200-W light bulb is brighter than the light from a 60-W light bulb. Does this mean that light coming from the 200-W bulb has a larger electric field? A larger magnetic field? A higher frequency? A longer wavelength?
4. Light from the Sun reaches Earth's atmosphere at a rate of about 1350 W/m^2. Assuming that this light is entirely absorbed, estimate the force exerted on Earth due to radiation pressure. Is this significant compared with the force on Earth due to the Sun's gravitational attraction?

Chapter Review

In this chapter, our understanding of the connection between electricity and magnetism culminates in the treatment of **electromagnetic waves**. As will be mentioned below, visible light is one example of an electromagnetic wave; these waves are sometimes referred to as *light* even when they cannot be seen with the naked eye (*electromagnetic radiation* is also a commonly used term).

25–1 The Production of Electromagnetic Waves

Electromagnetic waves are generated by accelerating charges. One of the most common ways of doing this is by connecting an antenna to an AC circuit. The charges accelerating back and forth in the antenna produce electromagnetic waves that travel away from the antenna at the speed of light. The electromagnetic wave consists of electric and magnetic fields that are perpendicular to one another and are in phase with one another. Electromagnetic waves are transverse waves; the direction of propagation is perpendicular to both $\vec{\mathbf{E}}$ and $\vec{\mathbf{B}}$. Given $\vec{\mathbf{E}}$ and $\vec{\mathbf{B}}$, the direction in which the wave propagates can be found from a right-hand rule:

> *Point the fingers of your right hand in the direction of* $\vec{\mathbf{E}}$ *so that they would curl toward* $\vec{\mathbf{B}}$. *Your thumb then gives the direction of propagation.*

Practice Quiz

1. Consider an electromagnetic wave for which the electric field points toward the top of this page, ↑, and the magnetic field points into this page, ⊗. What is the direction of propagation of the electromagnetic wave?

   (a) → **(b)** ⊙ **(c)** ↓ **(d)** ← **(e)** ⊗

25–2 The Propagation of Electromagnetic Waves

Unlike the other waves we have studied, electromagnetic waves do not require a medium; they can propagate in vacuum, because an electromagnetic wave is *self-sustaining* by electromagnetic induction. The changing magnetic field produces the changing electric field, and this changing electric field produces the changing magnetic field. The speed of an electromagnetic wave in vacuum is a fundamental constant of nature called the **speed of light**:

$$c = 3.00 \times 10^8 \text{ m/s}$$

From Maxwell's theory of electricity and magnetism, which predicted electromagnetic waves, we know that the speed of light in vacuum is related to the permittivity and permeability of free space,

$$c = \frac{1}{\sqrt{\varepsilon_0 \mu_0}}$$

The propagation of electromagnetic waves exhibits the Doppler effect similar to sound waves. For sound waves, the speed of the wave depends on the motion of the source; however, the speed of an electromagnetic wave is independent of the motion of the source. If the relative speed between the source and observer, u, is small compared to the speed of light, then the frequency received, f', is given by

$$f' = f\left(1 \pm \frac{u}{c}\right)$$

where f is the frequency emitted by the source. The + sign is used if the source is approaching the observer, and the – sign is used if the source is receding from the observer.

Example 25–1 The Doppler Effect Some frequencies of the light from another galaxy are found to be 0.65% lower than the corresponding frequencies from stationary sources on Earth. Is this galaxy moving toward or away from us? Determine the speed at which it is moving toward or away from us.

Picture the Problem The diagram shows the galaxy moving relative to Earth, either toward or away.

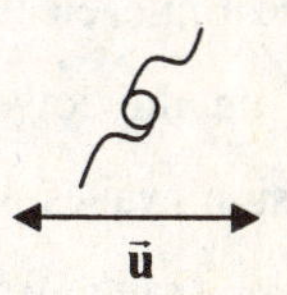

Strategy Given that the frequency is lower, we can deduce whether the galaxy is moving toward or away from us.

Solution

1. Because the frequency decreases, this implies: $\frac{f'}{f} < 1 \Rightarrow 1 \pm \frac{u}{c} < 1 \Rightarrow$ use – sign ∴ moving away

2. The observed frequency can be written as: $f' = f - 0.0065f = (0.9935)f$

3. The expression for the Doppler effect becomes: $\frac{f'}{f} = \left(1 - \frac{u}{c}\right) = \frac{(0.9935)f}{f} \quad \therefore \quad 1 - \frac{u}{c} = 0.9935$

4. Solving for u gives: $u = (0.0065)c = (0.0065)(3.00 \times 10^8\ \frac{\text{m}}{\text{s}}) = 2.0 \times 10^6\ \frac{\text{m}}{\text{s}}$

Insight This may look like an unreasonably high speed, but many distant galaxies are moving away from us very rapidly.

Practice Quiz

2. If an observer determines the frequency of the light given off by a source to be 10% higher than expected, what velocity, relative to the observer, must the source of the light have (in m/s)?

(a) 3.0×10^8 **(b)** 3.0×10^7 **(c)** 3.0×10^6 **(d)** 3.0×10^5 **(e)** 3.0×10^9

25–3 The Electromagnetic Spectrum

As with other types of waves, the speed of an electromagnetic wave equals the product of its wavelength and frequency:

$$c = \lambda f$$

The difference with electromagnetic waves is that the speed is constant. The above equation suggests that any combination of λf that equals c might be a valid electromagnetic wave — this is in fact the case. The full (infinite) range of frequencies (and wavelengths) is known as the **electromagnetic spectrum.**

The finite electromagnetic spectrum that is observed is divided into several regions. **Radio waves** make up the lowest-frequency region of practical importance. This region includes both radio and television waves in the frequency range of about 10^6 Hz – 10^9 Hz. Electromagnetic waves in the frequency range from about 10^9 Hz – 10^{12} Hz are called **microwaves.** Microwaves are commonly used for long-distance communication and cooking. **Infrared waves** fall in the frequency range of about 10^{12} Hz – 10^{14} Hz. The frequencies of infrared waves are just below that of red light. Most of the heat given off by common objects is in the form of infrared waves, and most handheld remote controls operate via infrared waves (or "IR" as it's often called). The part of the electromagnetic spectrum that we can see with our eyes is called the **visible light** region. The frequency of visible light is on the order of 10^{14} Hz. This light includes all the colors of the rainbow and the various mixtures that these colors can produce.

The frequency range just above that of visible light is **ultraviolet light**. The frequency range of this "UV" light is about 10^{15} Hz – 10^{17} Hz. Ultraviolet light from the sun can be harmful over time, but most harmful UV rays are blocked by Earth's ozone (O_3). This fact is a key reason why ozone depletion is an important current issue. **X-rays** fall in the frequency range of about 10^{17} Hz – 10^{20} Hz. These waves are very penetrating and widely used in medicine to "see" past skin and tissue. The last region of the electromagnetic spectrum is for frequencies above 10^{20} Hz; these waves are called **gamma rays.** These waves are given off by radioactive materials and are even more penetrating and damaging than X-rays.

Exercise 25–2 An Electromagnetic Wave A certain electromagnetic wave has a frequency of 5.11×10^{10} Hz. What type of radiation is it? What is its wavelength?

Solution: We are given that $f = 5.11 \times 10^{10}$ Hz.

In terms of the frequency, the wavelength is given by $\lambda = c/f$. Therefore,

$$\lambda = \frac{3.00 \times 10^{8} \text{ m/s}}{5.11 \times 10^{10} \text{ Hz}} = 5.87 \times 10^{-3} \text{ m}$$

The frequency of this wave falls within the range of microwaves.

Practice Quiz

3. A certain electromagnetic wave has a wavelength of 620 nm. What is its frequency?

(a) 4.8×10^{14} Hz **(b)** 2.1×10^{-15} Hz **(c)** 6.2×10^{-9} Hz **(d)** 6.2×10^{-7} Hz **(e)** 3.0×10^{8} Hz

4. A certain electromagnetic wave has a wavelength of 0.1 nm. What type of radiation is it?

(a) infrared **(b)** visible light **(c)** ultraviolet **(d)** X-rays **(e)** gamma rays

25–4 Energy and Momentum in Electromagnetic Waves

You may recall that energy is stored in both electric fields and magnetic fields. Therefore, electromagnetic waves carry energy. The energy density of an electromagnetic wave is the sum of the energy densities of the electric field and the magnetic field

$$u = u_E + u_B = \frac{1}{2}\varepsilon_0 E^2 + \frac{1}{2\mu_0}B^2$$

For an electromagnetic wave $u_E = u_B$, so that we can also write

$$u = \varepsilon_0 E^2 = \frac{B^2}{\mu_0}$$

The electric and magnetic fields of an electromagnetic wave vary sinusoidally with time, just like the voltage and current in an AC circuit. Therefore, the average energy density of the wave is conveniently written in terms of the rms values of the fields:

$$u_{av} = \frac{1}{2}\varepsilon_0 E_{rms}^2 + \frac{1}{2\mu_0}B_{rms}^2 = \varepsilon_0 E_{rms}^2 = \frac{B_{rms}^2}{\mu_0}$$

where $E_{rms} = E_{\max}/\sqrt{2}$, and $B_{rms} = B_{\max}/\sqrt{2}$. These results can also be used to show the direct relationship between the magnitudes of $\vec{\mathbf{E}}$ and $\vec{\mathbf{B}}$,

$$E = cB$$

As with sound waves, the intensity of electromagnetic waves is an important quantity. The intensity is the amount of energy delivered to a unit area in a unit time. In terms of the energy density, the intensity of electromagnetic waves is given by

$$I = uc = c\varepsilon_0 E^2 = cB^2 / \mu_0$$

The average intensity is obtained by using the rms values of E and B.

In addition to carrying energy, electromagnetic waves also carry momentum. The momentum transferred by an electromagnetic wave to an area that absorbs an amount of energy U is given by

$$p = \frac{U}{c}$$

This momentum transfer results in a force applied to the area. The force per unit area is the pressure; in the case of electromagnetic waves it is called **radiation pressure**. The average radiation pressure applied by waves with an average intensity I_{av} is

$$\text{Pressure}_{av} = \frac{I_{av}}{c}$$

Example 25–3 Energy Density The electric and magnetic fields of an electromagnetic wave are oriented as shown. **(a)** What is the direction of propagation of this wave? **(b)** If the average energy density of this wave is 3.28×10^{-6} J/m^3, what are the rms values of the electric and magnetic fields?

Picture the Problem The sketch shows the directions of $\vec{\mathbf{E}}$ and $\vec{\mathbf{B}}$ in an electromagnetic wave.

$\odot\ \vec{\mathbf{E}}$

$\downarrow\ \vec{\mathbf{B}}$

Strategy We can use the right-hand rule to get the direction of propagation. We can obtain the rms values of the $\vec{\mathbf{E}}$ and $\vec{\mathbf{B}}$ field by starting with the expression for the energy density.

Solution

Part (a) The right-hand rule for the direction of propagation gives:

$\longrightarrow \vec{\mathbf{c}}$

Part (b)

1. The average energy in terms of E_{rms} is:

$$u_{av} = \varepsilon_0 E_{rms}^2 \quad \Rightarrow \quad E_{rms} = \left(\frac{u_{av}}{\varepsilon_0}\right)^{1/2}$$

$$\therefore \quad E_{rms} = \left(\frac{3.28 \times 10^{-6}\ \text{J/m}^3}{8.85 \times 10^{-12}\ \text{C}^2/\text{N}\cdot\text{m}^2}\right)^{1/2} = 609\ \text{N/C}$$

2. Find the rms value of the magnetic field: $B_{rms} = \frac{E_{rms}}{c} = \frac{608.8 \text{ N/C}}{3.00 \times 10^8 \text{ m/s}} = 2.03 \times 10^{-6} \text{ T}$

Insight The rms magnetic field could also have been found from the energy density. Try it.

Example 25–4 Radiation Pressure For the electromagnetic wave in Example 25–3, what average force would it exert on a 1.00-cm × 1.00-cm slab?

Picture the Problem The sketch shows an electromagnetic wave impinging on a square slab.

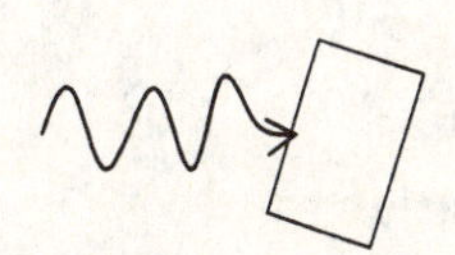

Strategy The pressure is force per area, so we can determine the force by using the radiation pressure.

Solution

1. The average radiation pressure P_{av} is: $P_{av} = \frac{I_{av}}{c} = \frac{u_{av}c}{c} = u_{av} = \frac{F_{av}}{A}$

2. Solving for the average force gives: $F_{av} = u_{av}A = \left(3.28 \times 10^{-6} \text{ J/m}^3\right)\left(1.00 \times 10^{-2} \text{ m}\right)^2 = 3.28 \times 10^{-10} \text{ N}$

Insight Note that the equations work out such that the average pressure equals the average energy density. Check that this makes sense dimensionally.

Practice Quiz

5. An electromagnetic wave whose electric field has an amplitude E_{max} contains an average energy density u_{av}. Another electromagnetic wave of amplitude $2E_{max}$ would contain an average energy density of

   (a) u_{av} (b) $\sqrt{2}\, u_{av}$ (c) $2u_{av}$ (d) $4u_{av}$ (e) $u_{av}/2$

6. The electric field of a certain electromagnetic wave has an rms value of 2130 N/C. What is the rms value of the associated magnetic field?

   (a) 1.4×10^5 T (b) 6.4×10^{11} T (c) 7.1×10^{-6} T (d) 3.0×10^8 T (e) 2130 T

7. The electric field of an electromagnetic wave has an rms value of 950 N/C. What is the average energy density of this wave (in J/m^3)?

(a) 4.0×10^{-6} **(b)** 2.4×10^{3} **(c)** 9.0×10^{5} **(d)** 8.4×10^{-9} **(e)** 8.0×10^{-6}

8. An electromagnetic wave whose electric field has an amplitude E_{max} exerts an average radiation pressure P_{av}. Another electromagnetic wave of amplitude $\sqrt{2}\,E_{max}$ would exert an average radiation pressure of

(a) $\sqrt{2}P_{av}$ **(b)** P_{av} **(c)** $2P_{av}$ **(d)** $4P_{av}$ **(e)** none of the above

25–5 Polarization

In general, the electric field in an electromagnetic wave points in all directions within a plane perpendicular to the direction of propagation; however, there are processes that can cause the electric field of an electromagnetic wave to oscillate with a specific orientation. Such electromagnetic waves are said to be **polarized**. When the electric field oscillates in random directions, the waves are called **unpolarized**. These same considerations apply to the magnetic field as well; however, by tradition, the direction of polarization is taken to be the direction of the electric field.

One of the ways in which electromagnetic waves are polarized is by passing through a **polarizer**. A polarizer is a material that absorbs electromagnetic waves whose electric fields are perpendicular to a certain direction called the **transmission axis**. Waves for which the electric field $\vec{\mathbf{E}}$ is parallel to this axis are transmitted through the material undiminished. If the electric field makes an angle θ with the transmission axis, the component of $\vec{\mathbf{E}}$ perpendicular to the axis is absorbed and the component parallel to the axis is transmitted. This leads to the **law of Malus**:

$$I = I_0 \cos^2\theta$$

where I_0 is the intensity of initially polarized waves whose polarization direction makes an angle θ with the transmission axis of the polarizer, and I is the intensity of the waves that emerge from the polarizer. The light that emerges from the polarizer will be polarized along the direction of the transmission axis. If the initial electromagnetic wave is unpolarized, the law of Malus shows that the intensity, upon passage through the polarizer, is reduced by half:

$$I = \tfrac{1}{2}I_0 \qquad \text{(initially unpolarized light)}$$

Often, a second polarizer, called an **analyzer**, is used to investigate the properties of polarized light, and if it's free to rotate, it will produce light of variable intensity. (Electromagnetic waves can also be polarized by scattering and reflection.)

Example 25–5 Polarization Unpolarized light of intensity 1.20×10^3 W/m^2 is incident on a polarizer-analyzer system. If the final intensity that emerges is 504 W/m^2, what is the angle between the transmission axes of the polarizer and the analyzer?

Picture the Problem The sketch shows an unpolarized electromagnetic wave incident on a polarizer-analyzer system.

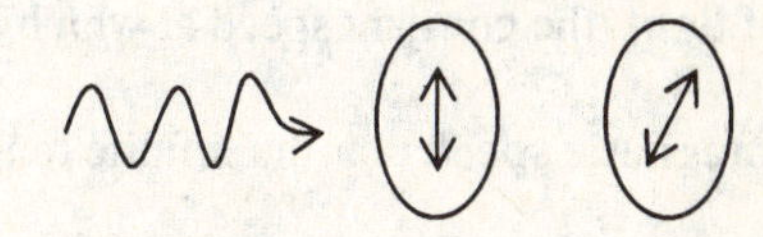

Strategy For the unpolarized light we use the result that the intensity is halved by the polarizer, and then we apply the law of Malus for the light that passes through the analyzer.

Solution

1. For the incident intensity I_0 and that which emerges from the polarizer, I_1, we have: $I_1 = \frac{1}{2} I_0$

2. Call the final intensity I_2, then from Malus' law we get: $I_2 = I_1 \cos^2\theta = \frac{1}{2} I_0 \cos^2\theta$

3. Solving for θ gives: $\theta = \cos^{-1}\left[\sqrt{\frac{2I_2}{I_0}}\right] = \cos^{-1}\left[\sqrt{\frac{2(504\ \text{W/m}^2)}{1.20\times10^3\ \text{W/m}^2}}\right] = 23.6^\circ$

Insight Here, we see that a system of polarizers is handled by applying the law of Malus to each polarizer.

Practice Quiz

9. If polarized light of intensity 750 W/m^2 is incident on a polarizer whose transmission axis makes an angle of 20.0° with the direction of polarization, what is the intensity of the light that emerges from the polarizer?

   (a) 750 W/m^2 **(b)** 375 W/m^2 **(c)** 705 W/m^2 **(d)** 662 W/m^2 **(e)** 0 W/m^2

10. If the intensity of initially polarized light is cut in half on passing through a polarizer, what is the angle between the direction of polarization and the transmission axis of the polarizer?

    (a) 0° **(b)** 30° **(c)** 45° **(d)** 60° **(e)** 90°

Reference Tools and Resources

I. Key Terms and Phrases

electromagnetic wave a wave of oscillating electric and magnetic fields

speed of light the constant speed at which light travels in vacuum

electromagnetic spectrum the infinite range of frequencies (and wavelengths) of electromagnetic waves

radiation pressure the pressure exerted by electromagnetic waves

polarized light an electromagnetic wave whose electric field oscillates along a specific direction

polarizer an object that absorbs the component of the electric field of an electromagnetic wave that is perpendicular to its transmission axis

transmission axis the direction in a polarizer that transmits a polarized electromagnetic wave without loss of intensity if it is polarized along this direction

law of Malus the expression for the intensity that results when polarized light passes through a polarizer

analyzer a second polarizer used to analyze the polarized light that emerged from a polarizer

II. Important Equations

Name/Topic	Equation	Explanation
Doppler effect	$f' = f\left(1 \pm \frac{u}{c}\right)$	The Doppler effect for light, assuming $u \ll c$
electromagnetic spectrum	$c = \lambda f$	The relationship between the wavelength and frequency of electromagnetic waves
energy density	$u_{av} = \frac{1}{2}\varepsilon_0 E_{rms}^2 + \frac{1}{2\mu_0}B_{rms}^2 = \varepsilon_0 E_{rms}^2 = \frac{B_{rms}^2}{\mu_0}$	The average energy density of electromagnetic waves
electromagnetic waves	$E = cB$	The relationship between E and B in an electromagnetic wave

intensity	$I = uc = c\varepsilon_0 E^2 = cB^2 / \mu_0$	The instantaneous intensity of electromagnetic waves
polarization	$I = I_0 \cos^2 \theta$	The law of Malus for the intensity of light emerging from a polarizer

Puzzle

LIGHT DRILLS

The sketch shows the basic elements used by Michelson to measure the speed of light. A pulse of light hits the rotating 8-sided mirror and travels to a flat mirror some distance away. The pulse will be received by the detector if the rotating mirror is in the right orientation when the pulse returns. A household power drill can turn at about 1200 rpm. If such a drill is used to turn the mirror, how far does the flat mirror have to be for a pulse to be observed? How about for a Dremel that turns at 30,000 rpm? (Hint: Assume the distance *L* is large compared to the size of the 8-sided mirror.)

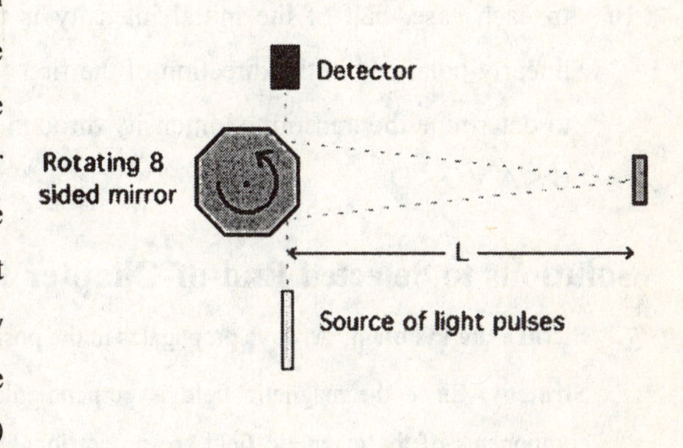

Answers to Selected Conceptual Questions and Exercises

Conceptual Questions

4. Light reflected from a horizontal surface has a polarization in the horizontal direction. It follows that when you sit upright, with the transmission axis of your glasses in the vertical direction, they will block most of the reflected light. When you lie on your side, however, the transmission axis is horizontal. This allows most of the reflected light to enter your eyes.

6. The light from the sky is polarized at right angles to the direction of the Sun; therefore, the amount of light received by each of the two polarizing eyes will depend on the orientation of the spider relative to the Sun. By monitoring the amount of light received by each eye, the spider can maintain a course on a given heading relative to the Sun.

10. As mentioned in the answer to Conceptual Question 4, the light reflected from a horizontal surface is polarized primarily in the horizontal direction. If the glasses are merely tinted, reflected light will have the same intensity no matter how the glasses are rotated. If they are Polaroid, however, you will notice a striking difference in reflected intensity as you rotate the glasses.

Conceptual Exercises

4. To find the direction of propagation of an E&M wave, point the fingers of the right hand in the direction of the electric field, curl them toward the direction of the magnetic field, and your thumb will point in the direction of propagation. Applying this rule, we find the following directions of propagation: Case 1, positive x direction; case 2, positive z direction; case 3, negative x direction.

6. Ideally, the sails should be reflecting. Recall that there is a greater transfer of momentum when a light beam (or a thrown ball, for that matter) is reflected than when it is merely absorbed.

10. In each case, half of the initial intensity is transmitted through the first filter. The light is then linearly polarized in the direction of the first filter. We then use the law of Malus (Equation 25-13) to determine the transmitted intensity through the second filter. The resulting ranking is as follows: $B < A = C < D$.

Solutions to Selected End-of-Chapter Problems

7. Picture the Problem: A wave propagates in the positive z-direction with a known magnetic field vector.

Strategy: Since the magnetic field is perpendicular to the electric field, the magnitudes of the x- and y-components of the magnetic field are proportional to the y- and x-components of the electric field, with the constant of proportionality given by Eq. 25-9. Use the right hand rule to determine the sign of each component.

Solution: 1. (a) Since $\vec{\mathbf{E}}$ is perpendicular to the direction of propagation, the z component is $\boxed{\text{zero}}$

2. (b) Switch the components of the magnetic field and multiply by c to calculate the magnitude of the electric field components:

$$\vec{\mathbf{E}} = c(-B_y\hat{\mathbf{x}} + B_x\hat{\mathbf{y}})$$

3. Insert the electric field components:

$$\vec{\mathbf{E}} = \left(3.00\times10^{8}\text{ m/s}\right)\left[-\left(-4.06\times10^{-9}\text{ T}\right)\hat{\mathbf{x}} + \left(2.32\times10^{-9}\text{ T}\right)\hat{\mathbf{y}}\right]$$
$$= \boxed{(1.22\text{ N/C})\hat{\mathbf{x}} + (0.696\text{ N/C})\hat{\mathbf{y}}}$$

Insight: Calculating the magnitude of the electric field gives $\sqrt{(1.22\text{ N/C})^2 + (0.696\text{ N/C})^2} = 1.40\text{ N/C}$ as stated in the problem.

13. Picture the Problem: Light from a galaxy is Doppler shifted to a frequency 15% lower than was emitted.

Strategy: Solve Eq. 25-3 for the speed of the galaxy with the frequency equal to 85% of the emitted frequency.

Solution: 1. (a) Due to the Doppler effect, the frequency of light as measured by an observer is less than what it was when emitted if the source is receding. So, the galaxy is moving $\boxed{\text{away}}$ from Earth.

2. (b) Set the shift frequency to 85% initial frequency:

$$f' = (1-0.15)f = 0.85f = f\left(1-\frac{u}{c}\right)$$

3. Solve for the speed:

$$0.85 = 1 - \frac{u}{c}$$
$$u = (1 - 0.85)c = \boxed{0.15c}$$

Insight: The percent change in frequency equals the percent of the speed of light that the galaxy is traveling.

33. Picture the Problem: As a wave enters a medium in which the speed of light decreases, the frequency remains constant, but the wavelength changes.

Strategy: Insert the new speed into Eq. 25-4 and write the new wavelength in terms of the old wavelength.

Solution: 1. (a) The wavelength of an electromagnetic wave is directly proportional to its speed. So, if the wave speed decreases, the wavelength $\boxed{\text{decreases}}$.

2. (b) Solve Equation 25-4 for the new wavelength:

$$\lambda' = \frac{\frac{3}{4}c}{f} = \frac{3}{4}\left(\frac{c}{f}\right) = \frac{3}{4}\lambda$$

3. Divide both sides by the initial wavelength:

$$\frac{\lambda'}{\lambda} = \boxed{\frac{3}{4}}$$

Insight: The speed of light in water is ¾ of the speed of light in a vacuum.

47. Picture the Problem: An electromagnetic wave has a 55 kW source and is transmitted isotropically.

Strategy: Since the wave is uniform in all directions, the intensity at any given point will equal the power divided by the area of the sphere whose radius is equal to the distance to the source. Calculate the intensity from the power divided by the area.

Solution: 1. (a) Calculate the intensity 250 m from the antenna:

$$I_{av} = \frac{P_{av}}{A} = \frac{P_{av}}{4\pi r^2} = \frac{55\times10^3\text{ W}}{4\pi(250\text{ m})^2} = \boxed{70\text{ mW/m}^2}$$

2. (b) Calculate the intensity 2500 m from the antenna:

$$I_{av} = \frac{55\times10^3\text{ W}}{4\pi(2500\text{ m})^2} = \boxed{0.70\text{ mW/m}^2}$$

Insight: Since the intensity is inversely proportional to the square of the distance, increasing the distance by a factor of 10 decreases the intensity by a factor of 100.

53. Picture the Problem: Sunlight of intensity 1.00 kW/m² is absorbed by a 15 m by 45 m black surface.

Strategy: Use Equation 25-12 to calculate the average pressure. Multiply the pressure by the area to calculate the average force.

Solution: 1. Calculate the average pressure:

$$P_{av} = \frac{I_{av}}{c} = \frac{(1.0\times10^3\text{ W/m}^2)}{3.00\times10^8\text{m/s}} = 3.33\mu\text{N/m}^2$$

2. Multiply the pressure by area:

$$F_{av} = P_{av}A = (3.33\times10^{-6}\text{ N/m}^2)(15\text{ m})(45\text{ m}) = \boxed{2.3\text{ mN}}$$

Insight: Compared with the other forces (gravity and air pressure), the force from the radiation is negligible.

69. Picture the Problem: Vertically polarized light of intensity 37.0 W/m^2 passes through two polarizers in the three orientations shown in the figure.

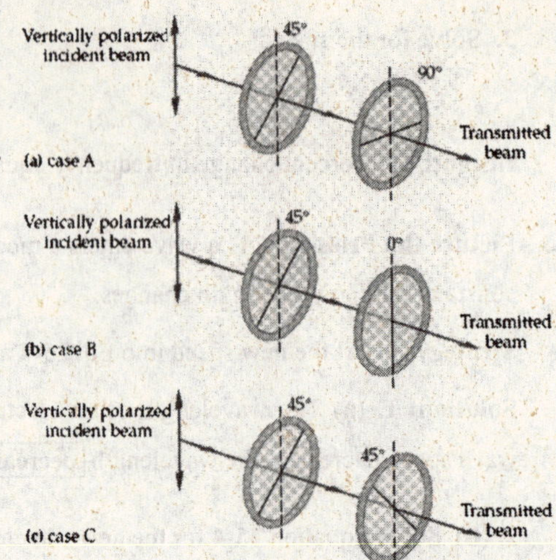

Strategy: Apply Malus's Law (Equation 25-13) for each of the lenses to determine the transmitted intensity.

Solution: 1. (a) In case A, both polarizers are at 45° angles to the beam polarization. The same it true for case B. Therefore the transmitted intensity for case A and B are the same. In case C the second polarizer is 90° to the beam polarization so the transmitted intensity is zero. Therefore, $\boxed{\text{(c) is the smallest; (a) and (b) tie.}}$

2. (b) Calculate the intensity for case A: $I = (37.0\text{ W/m}^2)\cos^2 45°\cos^2(90° - 45°) = \boxed{9.25\text{ W/m}^2}$

3. Calculate the intensity for case B: $I = (37.0\text{ W/m}^2)\cos^2 45°\cos^2(0° - 45°) = \boxed{9.25\text{ W/m}^2}$

4. Calculate the intensity for case C: $I = (37.0\text{ W/m}^2)\cos^2 45°\cos^2(45° + 45°) = \boxed{0}$

Insight: The relative magnitude of the intensities found in part (b) agree with the ordering given in part (a).

Answers to Practice Quiz

1. (d) **2.** (b) **3.** (a) **4.** (d) **5.** (d) **6.** (c) **7.** (e) **8.** (c) **9.** (d) **10.** (c)

CHAPTER 26
GEOMETRICAL OPTICS

Chapter Objectives

After studying this chapter, you should

1. know and be able to use the law of reflection.
2. be able to sketch ray diagrams to locate images formed by mirrors and lenses.
3. be able to use the mirror equation and the thin-lens equation together with their sign conventions.
4. be able to calculate the magnification of images formed by reflection and refraction.
5. be able to determine the speed of light in different media.
6. be able to use Snell's law for the refraction of light.
7. know how to determine the critical angle for total internal reflection and Brewster's angle.

Warm-Ups

1. In your own words, explain what a focal length is. Try not to use any equations or refer to any specific type of mirror or lens.
2. Estimate the focal length of a typical bathroom "magnifying mirror."
3. Images formed by spherical lenses can be real or virtual. What does it mean for an image to be virtual? Can such an image be seen?
4. Estimate the focal length of a typical magnifying glass.
5. What is the focal length of a clear, flat piece of glass, such as a normal household window?

Chapter Review

As discussed in the previous chapter, light is an electromagnetic wave; however, understanding the behavior of light does not always require a wave analysis. In this chapter and the next, we study the branch of physics, called **geometrical optics**, in which conditions are such that the wave nature of light can be "glossed over" while still accurately describing its behavior.

26–1 The Reflection of Light

In the study of geometrical optics, we think of light as **rays** traveling along a straight-line path. In terms of the propagating electromagnetic wave, we can track the crests of these waves (or any specific phase point). The collection of crests at a given phase can be imagined to form surfaces called **wave fronts**. Within this approximation, the light rays are perpendicular to the wave fronts and point in the direction of their propagation. As a light wave gets farther away from the source, the wave fronts are so spread out that the surfaces are approximately planar; these waves are called **plane waves** and they give rise to parallel rays.

The reflection of light from a smooth boundary obeys a simple law called the **law of reflection**. This law says that the angle the incident ray makes with the normal to the reflecting surface, the *angle of incidence*, θ_i, is equal to the angle that the reflected ray makes with the normal to the surface, called the *angle of reflection*, θ_r. The incident ray, the normal line, and the reflected ray all lie in the same plane.

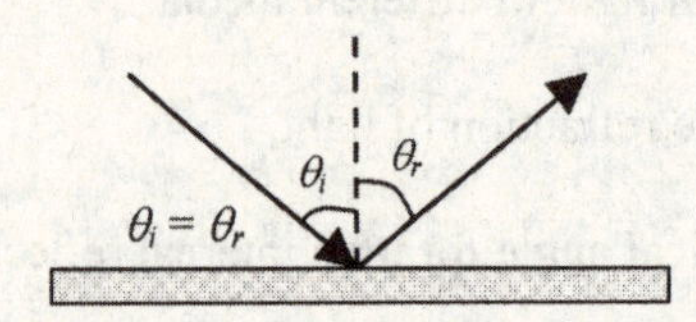

Practice Quiz

1. An incident light ray makes an angle of 33° with the normal to the surface of a plane mirror. What angle does the reflected ray make with the surface of the mirror?

   (a) 0° **(b)** 33° **(c)** 66° **(d)** 57° **(e)** none of the above

26–2 Forming Images with a Plane Mirror

A plane mirror is one that is perfectly flat. When you look at your reflection in a plane mirror, what you see is called an **image** of yourself. The source that is being reflected in the mirror (you) is called the **object**. Using the law of reflection, several results about reflection with a plane mirror can be found:

* The distance between the object and the mirror, called the **object distance**, d_o, is equal to the distance between the image, on the opposite (back) side of the mirror, and the mirror, called the **image distance**, d_i.

* The image is right side up, referred to as **upright**.

* The image is the same size as the object.

Left and right appear to be reversed in a mirror because the image is facing you. Just as when a person faces you, his or her right hand is on your left, and vice versa.

Practice Quiz

2. An object is 5.0 m in front of a plane mirror. How far away from the mirror is its image?

   (a) 0.0 m **(b)** 2.5 m **(c)** 5.0 m **(d)** 10.0 m **(e)** 15.0 m

26–3 Spherical Mirrors

A spherical mirror has the shape of a section of a sphere. If the reflecting surface is on the outside of this spherical section, it is called a **convex** mirror; if the reflecting surface is on the inside, it is called a **concave** mirror. The **principal axis** of the mirror is the line that passes through the center of the mirror (often called the *vertex*) perpendicular to the surface. A distance R away from the vertex along the principal axis is a point called the **center of curvature**, C, where R is the radius of the sphere from which the spherical section of mirror was taken (often called the *radius of curvature*). The point C would be located at the center of the sphere.

If light rays that are parallel to the principal axis and lie close to it (paraxial rays) are incident on a spherical mirror, they will either converge to (for a concave mirror) or diverge from (for a convex mirror) points at a distance of magnitude f from the vertex, called the **focal length**. The point on the principal axis a distance of one focal length away from the vertex is called the **focal point**. For a spherical mirror, the focal length is given by

$$f = \pm\tfrac{1}{2}R$$

where the + sign applies to concave mirrors, and the – sign applies to convex mirrors. The full reasoning behind this difference in sign will be discussed below. Basically, the positive sign means that the focal point is in front of the mirror, and the negative sign means that it is behind the mirror.

Practice Quiz

3. The center of curvature of a convex, spherical mirror is located 50 cm away from its vertex. What is its focal length?

   (a) 25 cm **(b)** –25 cm **(c)** 50 cm **(d)** –50 cm **(e)** none of the above

26–4 Ray Tracing and the Mirror Equation

Our main objective in image formation by reflection is to determine the size, location, and orientation of the image of a given object. There are two methods for obtaining this information, one method, **ray tracing**, is geometric, and the other method, the **mirror equation**, is algebraic. In practice, ray tracing is used qualitatively to help visualize the given situation, and the mirror equation is used to obtain accurate numerical results.

For ray tracing with spherical mirrors, there are three commonly used rays: the *P-ray* is drawn parallel to the principal axis, the *F-ray* (or its extension) intersects the principal axis at the focal point, and the *C-ray* (or its extension) intersects the principal axis at the center of curvature.

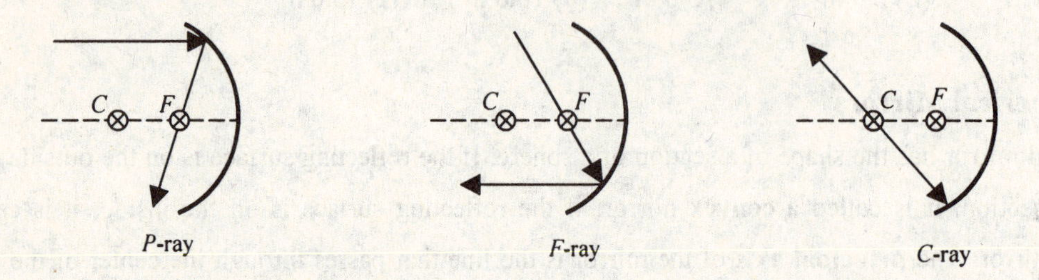

As you can see in the above figure for a concave mirror, the reflected *P*-ray intersects the principal axis at the focal point, the reflected *F*-ray is parallel to the principal axis, and the reflected *C*-ray hits the mirror at a 90 degree angle and travels back along the same path as the incident *C*-ray. When all three rays, coming from a given point on the object, are drawn, the intersection of the three reflected rays (or the intersection of their extensions) locates the image of that point.

In a version of the preceding diagram for a convex mirror, the incident *P*-, *F*-, and *C*-rays would be drawn similarly, except that the reflected rays would not intersect. Instead, the extensions of the reflected rays behind the mirror would intersect to form what is called a **virtual image** (see Figure 26-15 in the text). When the actual reflected light rays themselves intersect (not just their extensions), they form what is called a **real image**.

A geometrical analysis of a properly drawn ray diagram can be used to derive an equation that relates the image and object distances for reflection from a spherical mirror. The result of this analysis is the mirror equation

$$\frac{1}{d_o}+\frac{1}{d_i}=\frac{1}{f}=\frac{2}{R}$$

Similarly, an expression for the **magnification** of the image can be derived. The magnification is defined to be the ratio of the height of the image (h_i) to the height of the object (h_o). The result is

$$m=\frac{h_i}{h_o}=-\frac{d_i}{d_o}$$

In the ray-diagram approach, it will be clear from the diagram whether the image is real or virtual, upright or inverted. Algebraically, these attributes are distinguished by the sign that certain quantities have. Therefore, the mirror equation and the expression for the magnification are to be used in the context of certain *sign conventions* that distinguish among characteristics. The sign conventions for spherical mirrors are as follows:

Focal Length and Radius of Curvature

* f is positive for concave mirrors.
* f is negative for convex mirrors.

Magnification and Images

* m is positive for upright images.
* m is negative for inverted images.
* The image height is positive for upright images.
* The image height is negative for inverted images.

Image Distance

* d_i is positive for real images (those that are in front of the mirror).
* d_i is negative for virtual images (those that are behind the mirror).

Object Distance

* d_o is positive for real objects (those that are in front of the mirror).
* d_o is negative for virtual objects (those that are behind the mirror).

Example 26–1 A Convex Mirror A certain convex mirror has a radius of curvature of 75.0 cm. If a point source of light is placed 30.0 cm away from the mirror on the principal axis, where is the image of this object?

Picture the Problem The sketch shows a ray diagram for this situation. The diagram gives us some idea as to where the image should be.

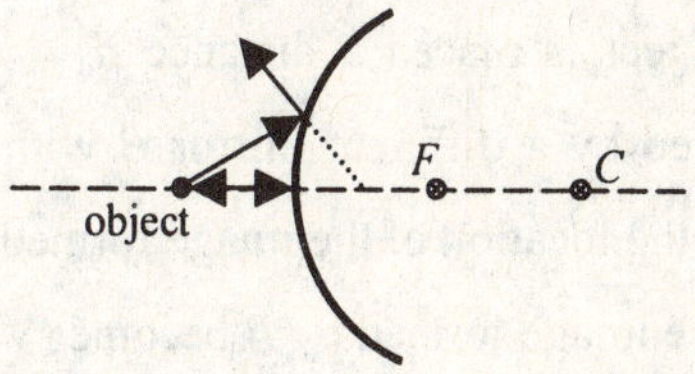

Strategy From the diagram, we expect a virtual image inside the focal point.

Solution

1. The values of f and d_o are: $f = -R/2 = -37.5$ cm, $d_o = 30.0$ cm

2. The mirror equation gives: $\dfrac{1}{d_i} + \dfrac{1}{d_o} = \dfrac{1}{f} \Rightarrow d_i = \left[\dfrac{1}{f} - \dfrac{1}{d_o}\right]^{-1}$

3. The numerical result is: $$d_i = \left[\frac{1}{-37.5\text{ cm}} - \frac{1}{30.0\text{ cm}}\right]^{-1} = -16.7\text{ cm}$$

Insight The final result is consistent with our expectation based on the ray diagram. Take note that the focal length is a negative number because it is a convex mirror. You must pay attention to the type of mirror being used. Because the object was a point source on the principal axis, the *P*-, *C*-, and *F*-rays were all the same, so I just drew an arbitrary ray as shown, and it still works.

Exercise 26–2 An Extended Object If the object in Example 26–1 was an extended object of height 1.50 cm, what would be the height and orientation of the image?

Solution

From the solution to Example 26–1 we know both the object and image distances. From the definition of the magnification we can write

$$\frac{h_i}{h_o} = -\frac{d_i}{d_o} \Rightarrow h_i = -\frac{h_o d_i}{d_o} = -\frac{(1.50\text{ cm})(-16.67\text{ cm})}{(30.0\text{ cm})} = 0.834\text{ cm}$$

Because this height is positive, we know that it is also an upright image.

Practice Quiz

4. If, for the reflection of a single object from a single spherical mirror we know that the image will be inverted, what else can we conclude about the image?

   (a) It is enlarged. **(b)** It is diminished. **(c)** It is real. **(d)** It is virtual. **(e)** none of the above.

5. An object is placed a distance $d_o > R$ away from a concave spherical mirror *A*. If the mirror is replaced by a different mirror *B* with a smaller radius of curvature, which of the following is true about the location of the image formed by *B* compared to that formed by *A*?

   (a) The image formed by *B* becomes virtual, whereas the image formed by *A* was real.

   (b) The image formed by *B* becomes real, whereas the image formed by *A* was virtual.

   (c) The image formed by *B* is closer to *B* than the image formed by *A* was to *A*.

   (d) The image formed by *B* is farther from *B* than the image formed by *A* was to *A*.

   (e) None of the above.

6. A concave spherical mirror has a focal length of 1.20 m. If an object is placed 75.0 cm in front of the mirror, where is its image?

   (a) 2.00 m in front of the mirror

   (b) 2.67 m behind the mirror

   (c) 0.500 m behind the mirror

   (d) 0.450 m in front of the mirror

   (e) None of the above.

26–5 The Refraction of Light

When light crosses the boundary between two different media, such as air and water, the light that penetrates into the second medium will, in general, travel in a different direction than the incident light; this phenomenon is known as **refraction.** Also, the speed of light in the second medium will generally be different from its speed in the first medium. The ratio of the speed of light in vacuum to its speed in a certain medium is called the **index of refraction** (n) of the medium:

$$n = \frac{c}{v}$$

Being the ratio of two speeds, the index of refraction is a dimensionless quantity. Note also that since c is the fastest known speed, $n \geq 1$. (See Table 26–2 in the text for values of n for different substances.)

Because the direction of the light ray changes upon transmission across the boundary, we require a way to determine the new direction of the ray. Let us call the angle that the incident ray makes with the normal line to the surface (in the first medium) θ_1 and the index of refraction of this medium n_1. The relationship between the angle of incidence and the angle, θ_2, that the refracted ray makes with the normal in the second medium (with index of refraction n_2) is

$$n_1 \sin\theta_1 = n_2 \sin\theta_2$$

This equation is called **Snell's law** (also known as the *law of refraction*). Some of the properties of refraction are illustrated in the following Example.

Example 26–3 Refraction Consider a smooth boundary between air and water. What is the angle of refraction if, **(a)** light is incident from the air with a 35° angle of incidence, **(b)** light is incident from the air perpendicular to the surface, and **(c)** light is incident from the water with a 35° angle of incidence?

Picture the Problem The diagram is for part (a) with the incident light ray in air and the refracted ray in water.

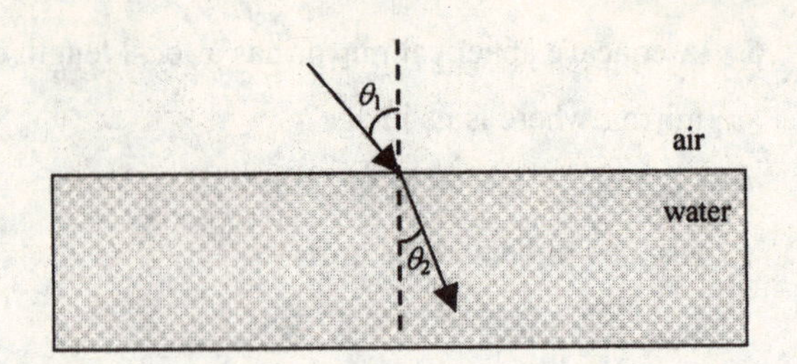

Strategy We can make direct use of Snell's law in each of the three cases.

Solution

Part (a)

1. Medium 1 is air and medium 2 is water. The values of n are from Table 26–2 in the text: $n_1 \sin\theta_1 = n_2 \sin\theta_2 \;\Rightarrow\; \theta_2 = \sin^{-1}\left[\dfrac{n_1 \sin\theta_1}{n_2}\right]$

2. The result is: $\theta_2 = \sin^{-1}\left[\dfrac{(1.00)\sin 35^\circ}{1.33}\right] = 26^\circ$

Part (b)

3. The angle of incidence is: $\theta_1 = 0.0^\circ$

4. Snell's law gives: $\theta_2 = \sin^{-1}\left[\dfrac{(1.00)\sin 0.0^\circ}{1.33}\right] = 0.0^\circ$

Part (c)

5. Medium 1 is water and medium 2 is air. Snell's law gives: $\theta_2 = \sin^{-1}\left[\dfrac{(1.33)\sin 35^\circ}{1.00}\right] = 50^\circ$

Insight Notice that in part (a) the refracted angle is smaller than 35°, and in part (c) it is bigger. These results reflect the general behavior that rays are refracted toward the normal if $n_2 > n_1$ and away from the normal if $n_2 < n_1$.

In general, when light is incident on a boundary there is both a reflected and a refracted ray. For a given pair of media, when $n_2 < n_1$, there is an angle of incidence beyond which no light is transmitted into the second medium; this angle is called the **critical angle** for **total internal reflection,** θ_c. The critical angle is reached when $\theta_2 = 90°$; therefore,

$$\sin\theta_c = \frac{n_2}{n_1}$$

Another effect that can be related to Snell's law is the total polarization of light that is reflected from a boundary. As mentioned in the previous chapter of the text, reflected light is generally partially polarized.

At a certain angle of incidence, called **Brewster's angle,** θ_B (sometimes also called the *polarization angle*), the reflected light is completely polarized parallel to the reflecting surface. Brewster's angle is the angle of incidence such that the angle between the reflected and refracted rays is 90°. Using this fact in Snell's law gives

$$\tan\theta_B = \frac{n_2}{n_1}$$

Practice Quiz

7. What is the speed of light in ice (in m/s)?

   (a) 3.00×10^8 **(b)** 1.81×10^8 **(c)** 2.00×10^8 **(d)** 2.26×10^8 **(e)** 2.29×10^8

8. For a light ray incident on a boundary from air (n = 1.00) to crown glass (n = 1.52), the refracted ray makes an angle of 27° with the normal. What is the angle of incidence?

   (a) 63° **(b)** 57° **(c)** 33° **(d)** 44° **(e)** 17°

9. Which of the following cases cannot produce total internal reflection? A light ray goes from

   (a) air to water
   (b) water to ice
   (c) flint glass to crown glass
   (d) diamond to water
   (e) none of the above

26–6 Ray Tracing for Lenses

The light rays bend upon transmission into a different medium; we can also achieve image formation by refraction. This way of forming images uses a **lens** instead of a mirror. Lenses come in many different types. Here, we consider a basic *converging lens* (double convex) and a basic *diverging lens* (double concave). The principal axis of the lens is the line passing through the center of the lens making right angles with the surfaces. Incident rays that are parallel to the principal axis will converge to a focus (for a converging lens) at a focal point F on the axis and at a local length f from the center of the lens.

As with mirrors, we can characterize the image formed by refraction with ray-tracing diagrams. Again, there are three commonly used rays: the *P-ray* is drawn parallel to the principal axis, the *F-ray* (or its extension) intersects the principal axis at the focal point, and the *M-ray* passes through the middle of the lens.

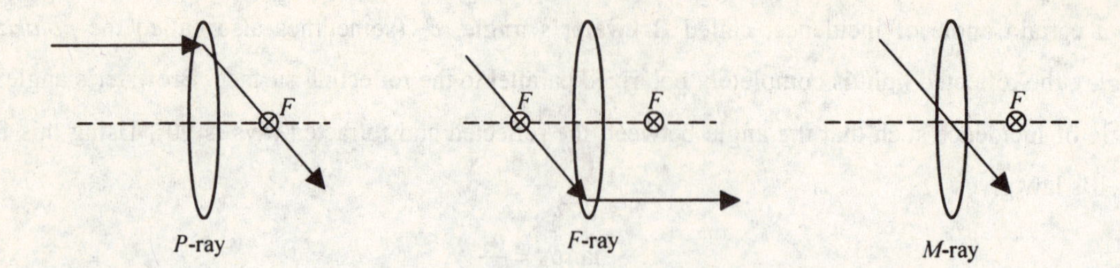

As you can see in the above figure for a (double) convex lens, the refracted *P*-ray intersects the principal axis at the focal point, the refracted *F*-ray is parallel to the principal axis, and the refracted *M*-ray passes through the center of the lens undeflected (which is a good approximation if the lens is thin). When all three rays, coming from a given point on the object are drawn, the intersection of the three refracted rays locates the image of that point.

In a version of the above diagram for a concave lens, the incident *P*-, *F*-, and *M*-rays would be drawn similarly, except that the refracted rays would not intersect. Instead, the extensions of the refracted rays back in front of the lens would intersect to form a **virtual image** (see Figure 26–34 in the text).

26–7 The Thin-Lens Equation

The geometry of a ray diagram for lenses can be used to derive a relationship between the image and object distances, as well as the magnifications. The results look just like those for spherical mirrors. The relationship between d_i and d_o is called the **thin-lens equation**:

$$\frac{1}{d_o}+\frac{1}{d_i}=\frac{1}{f}$$

The magnification of the image also obeys the same relationship that it does for mirrors:

$$m=\frac{h_i}{h_o}=-\frac{d_i}{d_o}$$

The sign conventions for using the thin-lens equation and the magnification are as follows:

Focal Length

* *f* is positive for converging lenses
* *f* is negative for diverging lenses

Magnification

* *m* is positive for upright images
* *m* is negative for inverted images

Image Distance

* d_i is positive for real images (those that are on the opposite side of the lens from the object)
* d_i is negative for virtual images (those that are on the same side of the lens as the object)

Object Distance

* d_o is positive for real objects (those from which light diverges)
* d_o is negative for virtual objects (those toward which light converges)

Example 26–4 The Image of a Concave Lens A concave lens has a focal length of 65.0 cm. The image of a certain object is virtual and has a magnification of 0.250. Determine both the image and object distances.

Picture the Problem The sketch is of a ray diagram for this problem.

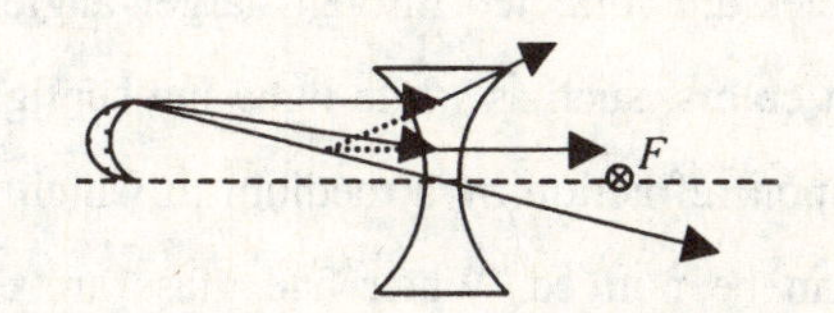

Strategy We have two equations that involve both unknowns, d_o and d_i, so we will solve them by substitution.

Solution

1. The magnification equation gives: $m = -\dfrac{d_i}{d_o} \Rightarrow d_i = -md_o$

2. Substitute this into the thin-lens equation: $\dfrac{1}{d_o} + \dfrac{1}{d_i} = \dfrac{1}{f} \Rightarrow \dfrac{1}{d_o} - \dfrac{1}{md_o} = \dfrac{1}{f} = \dfrac{m-1}{md_o}$

3. Solving for d_o gives: $d_o = \dfrac{f(m-1)}{m} = \dfrac{(-65.0\text{ cm})(0.250-1)}{0.250} = 195\text{ cm}$

4. Use the magnification to get d_i: $d_i = -md_o = -(0.250)(195\text{ cm}) = -48.8\text{ cm}$

Insight Notice again, as in Example 26–1 that we had to know to make the focal length negative because we were working with a diverging lens. If we had forgotten to do that, we would have got a positive image distance, indicating that the image was on the opposite side from what is indicated by the ray diagram. This discrepancy would have tipped us off to take a closer look at the solution. This consistency check will only work if you draw the ray diagrams, so do it when you solve these problems.

10. An object is 1.25 m away from a converging lens of focal length 95 cm. Which of the following correctly characterizes the image?

(a) virtual, enlarged, and inverted

(b) real, enlarged, and upright

(c) virtual, diminished, and upright

(d) real, enlarged, and inverted

(e) real, diminished, and inverted

26–8 Dispersion and the Rainbow

In addition to the fact that the index of refraction depends on the medium, for a given medium the index of refraction also depends on the frequency of the light that is being refracted. Generally, higher frequencies are refracted through larger angles. This means that light that is made up of a mixture of different colors, such as white light and sunlight, will be separated out, or dispersed, into those different colors upon refraction by a medium in which the variation of n with frequency is wide enough that the effect can be noticed. Water and glass are examples of such media. This phenomenon is known as **dispersion**, and it is responsible for the **rainbow** that we often see when sunlight passes through water droplets or a glass prism.

Reference Tools and Resources

I. Key Terms and Phrases

geometrical optics the study of the reflection and refraction of light under the ray approximation

light ray a representation of the straight-line propagation of light

law of reflection the fact that the angle of incidence equals the angle of reflection for light reflecting from a smooth surface

object the source of the light from which an image will be formed by a mirror or a lens

image the representation of an object resulting from the convergence, or apparent convergence, of light rays coming from the object by reflection or refraction

upright image an image that is right side up with respect to the object

inverted image an image that is upside down with respect to the object

magnification ratio of the image height to the object height

focal point the point on the principal axis at which incident rays that are parallel to the principal axis (or their extensions) will intersect after reflection or refraction

focal length the distance of the focal point from the mirror or lens

real image an image formed by the intersection of light rays

virtual image an image formed by the intersection of extensions of light rays into a region where no light is actually present

refraction the phenomenon that the direction of propagation of light generally changes when it crosses the boundary between two media

index of refraction the ratio of the speed of light in vacuum to the speed of light in a medium

Snell's law relates the angle of incidence to the angle of refraction

total internal reflection a phenomenon that occurs at incident angles greater than that for which the angle of refraction is 90°

Brewster's angle the angle at which reflected light is completely polarized

dispersion the spreading of light of different frequencies owing to the fact that the index of refraction is frequency dependent

II. Important Equations

Name/Topic	Equation	Explanation
reflection	$\theta_i = \theta_r$	The law of reflection
spherical mirrors	$\frac{1}{d_o} + \frac{1}{d_i} = \frac{1}{f} = \frac{2}{R}$	The mirror equation
magnification	$m = \frac{h_i}{h_o} = -\frac{d_i}{d_o}$	This expression for the magnification applies to both mirrors and lens
refraction	$n = \frac{c}{v}$	The definition of the index of refraction
refraction	$n_1 \sin\theta_1 = n_2 \sin\theta_2$	Snell's law, which describes the refraction of light

thin lenses	$\frac{1}{d_o}+\frac{1}{d_i}=\frac{1}{f}$	The thin-lens equation

III. Know Your Units

Quantity	Dimension	SI Unit
magnification (m)	dimensionless	—
index of refraction (n)	dimensionless	—

Puzzle

PHYSICS LITE

The figure shows a concave mirror with f_M = 12.5 cm and a converging lens with f_L = 25 cm. They are placed 50 cm apart with an object centered between them. The lens will create images of both the object and its reflection. Describe the size and location of the images formed.

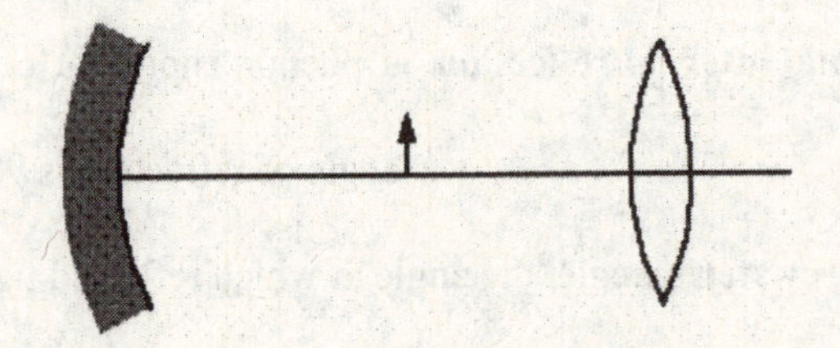

Bonus question: Describe how this setup could be useful in a lighthouse. Answer this question in words, not equations, briefly explaining how you obtained your answer.

Answers to Selected Conceptual Questions and Exercises

Conceptual Questions

4. The concave side of the dish collects the parallel rays coming from a geosynchronous satellite and focuses them at the focal point of the dish. The convex side of the dish would send the parallel rays outward on divergent paths. The situation is analogous to that of light in an optical telescope.

8. You are actually seeing light from the sky, which has been bent upward by refraction in the low-density air near the hot ground. See Figure 26-23 for a case where one would see a tree reflected in the "pool of water."

12. The oil used in the bottle to the left has an index of refraction that is equal to the index of refraction of the glass in the eye dropper. Therefore, light is undeflected when it passes from the oil to the glass or from the glass to the oil. Since light propagates the same as if the eye dropper were not present, the dropper is invisible.

14. Note that the word "SECRET", which is in red, is inverted. On the other hand, the word "CODE", which is in blue, appears not to be inverted. One might think that the different index of refraction for blue light versus red light is responsible for this behavior, but recall that we said the word "CODE" *appears* not to be inverted. In fact, it is inverted, just like the word "SECRET", but because all of its letters have a vertical symmetry, it looks the same when inverted.

Conceptual Exercises

6. Referring to Figure 26-18 (a), we see that as the object is moved farther to the left, the image moves to the right – toward the focal point of the mirror.

12. When light goes from air to glass it slows down; when it goes from glass to air it speeds up. In general, the speed of light is determined solely by the medium in which it propagates, irrespective of its past history.

18. **(a)** Over the period of a year on an airless Earth, the number of daylight hours would be equal to the number of nighttime hours. **(b)** Because the Earth's atmosphere bends light toward the normal as it enters the atmosphere – along a more vertical direction – we see the Sun as being higher in the sky than it actually is. As a result, we see the Sun even after a straight line to the Sun is below the horizon – therefore, the number of daylight hours, when averaged over a year, is greater than the number of nighttime hours. In fact, the atmosphere gives us more than an extra 24 hours of daylight over the course of a year.

Solutions to Selected End-of-Chapter Problems

7. Picture the Problem: The image shows you observing your belt buckle in a small vertical mirror hanging on a wall 2.00 meters in front of you. The height of the mirror is a distance h below your eye level.

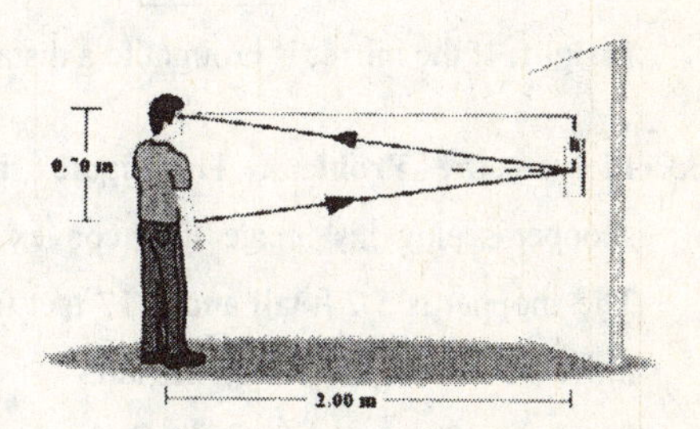

Strategy: Since $\theta_i = \theta_r$, the mirror must be midway between your eyes and belt buckle. Calculate the angle of reflection (the angle you must look down to see your belt buckle) from the inverse tangent of the ratio of the mirror height and distance.

Solution: 1. (a) Calculate the height of the mirror: $h = d/2 = (0.70\text{ m})/2 = \boxed{35\text{ cm}}$

2. (b) Calculate the reflected angle: $\theta = \tan^{-1}\dfrac{0.35\text{ m}}{2.0\text{ m}} = \boxed{9.9°}$

3. (c) Since the vertical position of the mirror relative to your eyes is halfway between your eyes and belt buckle regardless of the distance you stand from the mirror, $\boxed{\text{you will still see the buckle}}$.

Insight: Although you will always see your belt buckle, the angle at which you see the mirror will decrease as you walk away from the wall.

13. Picture the Problem: A person looks through a mirror of height 0.32 m while standing 95 meters from a building. He is holding the mirror 0.50 m in front of his eye.

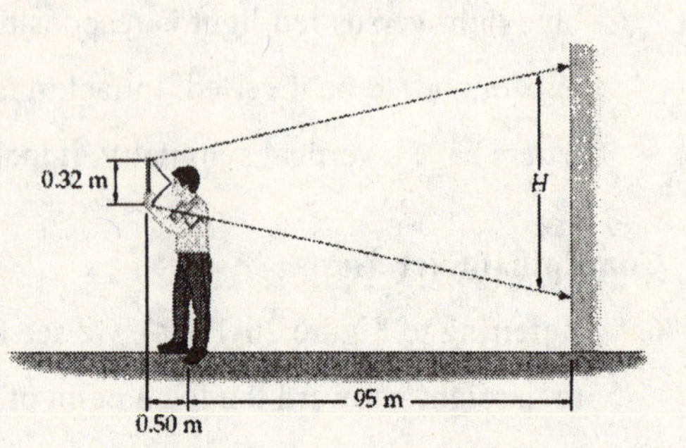

Strategy: Since the incident angle and the reflected angle are equal, the light rays from the ends of the window will intersect at the same distance behind the mirror as the eye is in front of the mirror. Using this relationship, create two similar triangles to relate the ratio of the height of the mirror to the mirror-eye distance and the ratio of the visible height of the building to the distance from the eye to the building. The distance from the eye to the building is equal to the sum of the distance from the building to you and twice the distance from you to the mirror. Set the two ratios equal and solve for the visible height of the building.

Solution: 1. (a) Set the ratios of the heights to distances equal: $\dfrac{h_{\text{mirror}}}{x_{\text{m-e}}} = \dfrac{H}{2x_{\text{m-e}} + x_{\text{e-b}}}$

2. Solve for the visible height of the building: $H = \dfrac{h_{\text{mirror}}\left(2x_{\text{m-e}} + x_{\text{m-b}}\right)}{x_{\text{m-e}}} = \dfrac{0.32\text{ m}\left(2\times 0.50\text{ m} + 95\text{ m}\right)}{0.50\text{ m}} = \boxed{61\text{ m}}$

3. (b) Decreasing $x_{\text{m-e}}$ increases the ratio $h_{\text{mirror}}/x_{\text{m-e}}$ and thus increases H. So, if the mirror is moved closer, the answer to part (a) will $\boxed{\text{increase}}$.

Insight: If the mirror is brought to a distance of 0.25 m from the eye, 122 m of the building would be visible.

33. Picture the Problem: The figure shows a shopper seeing his image in a convex mirror. The shopper is 5.7 ft tall and is 17 feet from the mirror. The image is 6.4 inches tall.

Shopper

Image

Strategy: Solve Equation 26-8 for the image distance. Then solve Equation 26-6 for the mirror's focal length. Finally use Equation 26-2 to calculate the radius of curvature.

Solution: 1. (a) Convex mirrors always produce upright and reduced images; the shopper's image is $\boxed{\text{upright}}$.

2. (b) Calculate the image distance: $d_i = -\frac{h_i}{h_o}d_o = -\left(\frac{6.4 \text{ in}}{5.7 \text{ ft}}\right)17 \text{ ft} = -19 \text{ in}\left(\frac{1 \text{ ft}}{12 \text{ in}}\right) = -1.59 \text{ ft}$

3. Calculate the focal length: $f = \left(\frac{1}{d_o}+\frac{1}{d_i}\right)^{-1} = \left(\frac{1}{17 \text{ ft}}+\frac{1}{-1.59 \text{ ft}}\right)^{-1} = -1.75 \text{ ft}$

4. Calculate the radius of curvature: $R = -2\left(\frac{1}{d_o}+\frac{1}{d_i}\right)^{-1} = -2f = -2(-1.75 \text{ ft}) = \boxed{3.5 \text{ ft}}$

Insight: If the mirror had been concave instead of convex, the shopper would have seen a real, inverted image of himself 1.95 feet in front of the mirror.

51. Picture the Problem: The image shows a beam incident upon the horizontal glass surface. This beam refracts into the glass. When the refracted light hits the vertical surface it totally internally reflects.

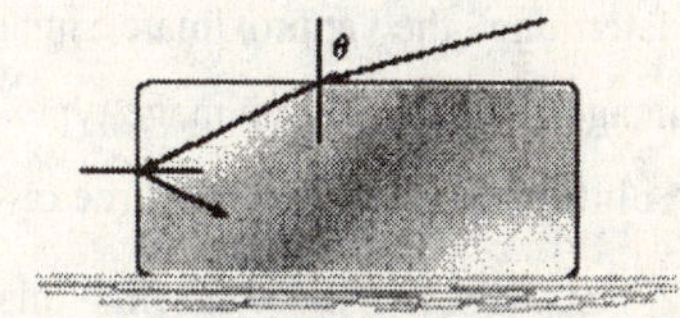

Strategy: From the figure, note that $\sin\theta_c = \sin(90° - \theta_r) = \cos\theta_r$. Use Snell's Law (Equation 26-11) to write an equation relating the index of refraction and the refracted angle. Use Equation 26-12 to write a second equation relating the refracted angle and index of refraction, where the sine of the critical angle is the cosine of the refracted angle. Square both of these equations and sum them to eliminate the angle and solve for the index of refraction.

Solution: 1. (a) Write Snell's Law: $n \sin\theta_r = \sin\theta_i$

2. Write Equation 26-12 in terms of the refracted angle: $\sin\theta_c = 1/n = \cos\theta_r$

$n\cos\theta_r = 1$

3. Sum the squares of the two equations and solve for n:

$n^2\sin^2\theta_r + n^2\cos^2\theta_r = \sin^2\theta_i + 1$

$n^2\left(\sin^2\theta_r + \cos^2\theta_r\right) = n^2 = \sin^2\theta_i + 1$

$n = \sqrt{\sin^2\theta_i + 1} = \sqrt{\sin^2 70° + 1} = \boxed{1.4}$

4. (b) Since the minimum index of refraction is related to the incident angle as $n = \sqrt{\sin^2\theta_i + 1}$, decreasing θ_i will cause n to be $\boxed{\text{decreased}}$.

Insight: As an example of decreasing the incident angle, set the incident angle equal to 50°. In this case the minimum index of refraction drops to 1.3.

61. Picture the Problem: The figure shows a convex lens ($f = 14$ cm) and a concave lens ($f = -7.0$ cm) separated by 35 cm. An object is located 24 cm in front of the convex lens.

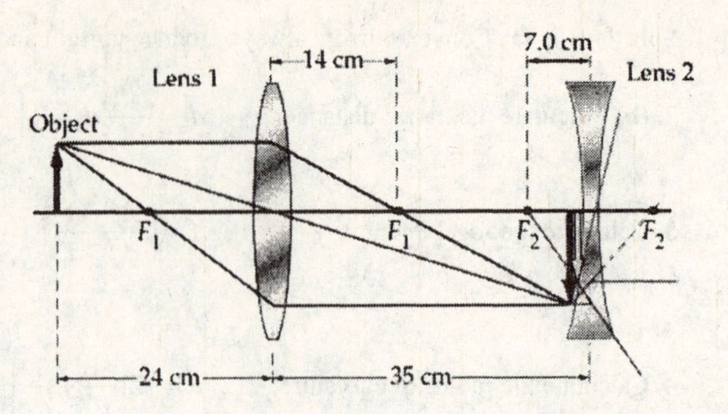

Strategy: Sketch the P-ray, F-ray, and M-ray for light from the object as it passes through the convex lens. From the rays draw the location of the image. Then treat the image as the object for the second lens and sketch the P-ray, F-ray, and M-ray for light as it passes through the concave lens. From the resulting graph, determine the approximate image position, image orientation, and image type.

Solution 1. (a) Sketch the three rays on the diagram:

2. Note the location of the final image: The image is located just to the left of Lens 2.

3. (b) Note the final image orientation: The image is inverted.

4. (c) Note the final image type: Since the final image is on the same side of Lens 2 as its object (the original image, which is real), it is virtual.

Insight: For any combination of lenses, the image from each lens becomes the object for the next lens.

Answers to Practice Quiz

1. (d) **2.** (c) **3.** (b) **4.** (c) **5.** (c) **6.** (e) **7.** (e) **8.** (d) **9.** (a) **10.** (d)

CHAPTER 27

OPTICAL INSTRUMENTS

Chapter Objectives

After studying this chapter, you should

1. know the basics of how the human eye works as an optical system.
2. understand the basics of how a camera works.
3. be able to characterize the image of multiple-lens systems.
4. understand the optics of correcting eyesight.
5. understand how magnifying glasses, microscopes, and telescopes produce enlarged images.
6. know what is meant by lens aberrations.

Warm-Ups

1. Nearsighted people can see objects clearly if they are close, but not if they are far away. Does a nearsighted eye have a focal length that is too long or too short?
2. Reading glasses are supposed to give a virtual image of an object held at arm's length farther out, at a point where the farsighted person can see clearly. Estimate the focal length of a reading-glass lens such that a newspaper held at 25 cm from the eyes will image at 40 cm from the eye.
3. In a simple refracting telescope there are two lenses, the objective lens and the eyepiece. Explain what each of these lenses does.
4. You wish to make a telescope using lenses that you have around. Let's say you have a magnifying glass and a pair of reading glasses. Estimate the highest magnification you can get.

Chapter Review

Geometrical optics, discussed in the last chapter, has many important applications in science and everyday life. This chapter discusses several of these applications.

27–1 The Human Eye and the Camera

The Human Eye

The human eye is the most important optical instrument we have. The outer coating of the eye is called the **cornea**. Most of the refraction of light rays in the eye occurs at the cornea. After passing through the cornea, light enters the eye through the **pupil**. The amount of light that passes through the pupil depends on its size, which is controlled by the **iris**. The iris expands in the dark and contracts in bright light to adjust the amount of light entering the eye. After entering the eye, light passes through an adjustable **lens**. In a process known as **accommodation**, the shape of the lens is adjusted by the **ciliary muscles** to help focus light onto the **retina** at the back of the eye.

In characterizing a person's vision, there are two important distances, the **near point** and the **far point**. The near point, N, is the closest distance to the eye that an object can be placed and still be in focus. For young people with good vision, the near point is typically about 25 cm. Since light rays from nearby objects require greater refraction, the ciliary muscles contract the lens as much as possible for objects at the near point, and the eye is under maximum *strain*. The far point is the greatest distance from the eye at which an object can be placed and still be in focus. For people with normal vision, the far point is effectively infinity. The light rays of distant objects require little refraction, so the ciliary muscles don't need to contract the lens for objects at the far point, and the eye is *relaxed* in this case.

The Camera

A camera and the human eye operate similarly. Light enters a camera through its **aperture**. The amount of light that passes through the aperture depends on its area, which is controlled by the ***f*-number** setting. The f-number equals the ratio of the focal length of the lens, f, to the diameter of the aperture, D,

$$f\text{-number} = \frac{f}{D}$$

Because the aperture is circular, its area, and therefore the amount of light that passes through it, is proportional to the square of the diameter, D^2.

Another feature of a camera that controls the amount of light that enters it is a setting called the **shutter speed**. This setting refers to the amount of time that the shutter (which blocks light from entering the camera) is open. A shutter speed setting of 1000, for example, means that the shutter speed is 1/1000th of a second, which is how long the shutter is open. Thus, the total amount of light that enters a camera depends on the combination of the settings for the f-number and the shutter speed.

Exercise 27–1 Camera Settings Suppose that for photograph 1 a camera is set at $f/2.8$ and a shutter speed setting of 500, and for photograph 2 the settings are $f/5.6$ and a shutter speed setting of 250. Which settings let in more light to the camera and by what percentage?

Solution

Let f_1 represent the *f*-number for photograph 1 (with D_1 as the associated aperture diameter) and f_2 for photograph 2 (with D_2 defined similarly to D_1). Since *f*-number $= f/D$, the ratio of the *f*-numbers is

$$\frac{f_2}{f_1} = \frac{f/D_2}{f/D_1} = \frac{D_1}{D_2} = \frac{5.6}{2.8} = 2$$

Thus, D_1 is 2 times the size of D_2, and since the amount of light passing through the aperture depends on the area (the square of the diameter), the *f*-number setting for photograph 1 lets in 4 times the amount of light as the *f*-number setting for photograph 2.

Now, considering the shutter speed, the speed for photograph 1 is 1/500 s and that for photograph 2 is 1/250 s; therefore, the shutter speed setting for photograph 1 lets in half the light as that for photograph 2. Combining these two results, we see that the settings for photograph 1 let in $4/2 = 2$ times more light than the settings for photograph 2. Hence, the settings are such that photograph 1 is taken with 100% more light than photograph 2 is.

The important thing to remember with camera settings is that you must consider both the *f*-number and the shutter speed settings when trying to assess the amount of light used for a photograph.

Practice Quiz

1. Which of the following is not part of the human eye?

(a) cornea **(b)** pupil **(c)** iris **(d)** near point **(e)** lens

2. Which of the following *f*-number settings represents the largest aperture diameter?

(a) 2.8 **(b)** 4 **(c)** 5.6 **(d)** 11 **(e)** 16

3. Which of the following shutter speed settings represents the shortest amount of time for the shutter to be open?

(a) 60 **(b)** 125 **(c)** 250 **(d)** 500 **(e)** 1000

27–2 Lenses in Combination and Corrective Optics

Optical systems involving more than one lens can be understood by noting two basic optical principles:

* The image produced by one lens acts as the object for the next lens in the system
* The total magnification produced by a system of lenses equals the product of the magnifications produced by each individual lens.

These principles apply to many optical systems; in this section, we use them to describe the corrective optics for eyesight.

Nearsightedness

The condition known as **nearsightedness**, or *myopia*, refers to the sight of a person who is able to focus nearby objects, but cannot focus very distant objects. For such a person, the far point is a finite distance from the eye (instead of infinity, as for normal vision). With nearsightedness, light rays are refracted too much, causing the image to form in front of the retina. This condition is corrected by placing a diverging lens in front of the eye. When a diverging lens with the correct **refractive power** is used, it forms a virtual image of very distant objects ($d_o = \infty$) at the person's far point, which enables him/her to focus this object.

The refractive power of a lens is the inverse of its focal length:

$$\text{refractive power} = \frac{1}{f}$$

When the focal length is measured in meters, we get the SI unit of refractive power, called a **diopter**, where 1 diopter = 1 m^{-1}.

Farsightedness

The condition known as **farsightedness**, or *hyperopia*, refers to the sight of a person who is able to focus very distant objects, but cannot focus nearby objects. For such a person, the near point is farther away from the eye as compared with normal vision. With farsightedness, light rays are not refracted enough, causing the image to form behind the retina. This condition is corrected by placing a converging lens in front of the eye. When a converging lens with the correct refractive power is used, it forms a virtual image of objects closer than the person's near point at locations beyond this near point, which means that the person's eye can now focus these objects. Typically, the corrective lens is chosen such that the virtual image of objects placed at the normal near point of 25 cm from the eye will be located at the person's actual near point.

Example 27–2 A Two-lens System An object is placed 75.0 cm in front of a converging lens of focal length $f_1 = 15.0$ cm. A second lens of focal length 10.0 cm is located 12.0 cm from the first lens on the opposite side from the object. Determine the location and magnification of the image produced by this system.

Picture the Problem The sketch shows the object as an arrow in front of the system of two lenses.

object lens 1 lens 2

Strategy We follow the basic strategy of making the image formed by the first lens the object for the second lens.

Solution

1. The object distance, d_{o1}, for the first lens is 75.0 cm; use the thin-lens equation to get the image distance d_{i1}:

$$\frac{1}{d_{i1}} + \frac{1}{d_{o1}} = \frac{1}{f_1} \Rightarrow d_{i1} = \left[\frac{1}{f_1} - \frac{1}{d_{o1}}\right]^{-1}$$

$$d_{i1} = \left[\frac{1}{15.0\text{ cm}} - \frac{1}{75.0\text{ cm}}\right]^{-1} = \frac{75.0\text{ cm}}{4} = 18.75\text{ cm}$$

2. Because the image from lens 1 is beyond lens 2, it is a virtual object ($d_{o2} < 0$) for lens 2. The object distance is:

$$d_{o2} = -(18.75\text{ cm} - 12.0\text{ cm}) = -6.75\text{ cm}$$

3. Using the thin-lens equation for the second lens, we can get the image distance d_{i2}:

$$d_{i2} = \left[\frac{1}{f_2} - \frac{1}{d_{o2}}\right]^{-1} = \left[\frac{1}{10.0\text{ cm}} - \frac{1}{-6.75\text{ cm}}\right]^{-1} = 4.03\text{ cm}$$

4. The location of the image is:

4.03 cm behind lens 2

5. The magnification is the product of the individual magnifications:

$$M = \left(-\frac{d_{i1}}{d_{o1}}\right)\left(-\frac{d_{i2}}{d_{o2}}\right) = \frac{d_{i1}d_{i2}}{d_{o1}d_{o2}}$$

$$= \frac{(18.75\text{ cm})(4.0299\text{ cm})}{(75.0\text{ cm})(-6.75\text{ cm})} = -0.149$$

Insight The two things to be careful of in these situations are (a) to make sure that you account for the separation between the lenses and (b) to watch for virtual objects and to use their object distances as negative values. In the problem, notice that the final image is real and inverted.

Example 27–3 A Nearsighted Correction The left eye of a person has a far point of 3.50 m. Assuming that the lens of her glasses sits 2.00 cm from her eye, what refractive power should be used for a corrective lens?

Picture the Problem The sketch shows a human eye with a corrective diverging lens in front of it.

Strategy To correct nearsightedness, the diverging lens forms a virtual image of objects infinitely far away.

Solution

1. The object distance of the objects we need to consider is infinite: $d_o = \infty \Rightarrow \dfrac{1}{d_o} = 0$

2. The image distance from the lens is: $d_i = 3.50\text{ m} - 2.00\text{ cm} = 3.48\text{ m}$

3. Using the thin-lens equation gives: $\dfrac{1}{f} = \dfrac{1}{d_i} + 0 \Rightarrow \dfrac{1}{f} = \dfrac{1}{d_i} = \dfrac{1}{-3.48\text{ m}} = -0.287\text{ diopter}$

Insight In finding the object distance, it didn't matter that the lens was 2.00 cm from the eye because we were considering objects at infinity. It was needed, however, to account for this distance with regard to the image distance and to remember that the image is virtual, so the image distance must be negative.

Practice Quiz

4. If the refractive power of a lens is 0.250 diopter, what is the focal length of a lens with twice the refractive power?

   (a) 0.500 m **(b)** 2.00 m **(c)** 4.00 m **(d)** 8.00 m **(e)** 0.125 m

5. A person whose near point is 28 cm and whose far point is 85 cm would most likely be considered

   (a) farsighted **(b)** hyperopic **(c)** nearsighted **(d)** all of the above **(e)** none of the above

27–3 The Magnifying Glass

A **magnifying glass** is a converging lens, which is used in much the same way as it is used for the correction of farsightedness. The magnifying glass allows you to focus objects that are closer to your eye than your near point by forming a virtual image of the object at infinity (if you place the magnifying glass one focal length away from the object). Because the object is now closer to your eye, it appears larger to you. Without the magnifying glass, the focused object would appear largest to you, having an angular size θ, if placed at your near point N. The **angular magnification**, M, of the object is the ratio of the angular size of the magnified object, θ', to θ. It can be shown that this ratio also equals the ratio of the near point, N, to the focal length of the lens f,

$$M = \frac{\theta'}{\theta} = \frac{N}{f}$$

Practice Quiz

6. A magnifying glass of focal length f produces an angular magnification of M. All else being equal, what would be the angular magnification if the lens was replaced by one of focal length $3f$?

(a) M **(b)** $3M$ **(c)** $6M$ **(d)** $M/3$ **(e)** $M/6$

27–4 The Compound Microscope

A **microscope** is an optical instrument designed to magnify small objects. The simplest microscope, called a *compound microscope*, consists of just two lenses. One lens, called the **objective**, is placed such that the object is just beyond its focal point. The objective forms a real, enlarged image at the focal point of the second lens, called the **eyepiece**. Essentially, the eyepiece acts as a magnifying glass that forms a virtual image at infinity. To a good approximation, the angular magnification of the compound microscope is given by

$$M = -\frac{d_i N}{f_o f_e}$$

where d_i is the image distance for the image formed by the objective, N is the near point of the observer, f_o is the focal length of the objective, and f_e is the focal length of the eyepiece.

Practice Quiz

7. If the objective of a microscope is moved farther away from the object, what effect does this have on the size of the image?

(a) It is enlarged.

(b) It is diminished.

(c) It becomes real from having been virtual.

(d) It becomes virtual from having been real.

(e) None of the above.

27–5 Telescopes

A telescope is an optical instrument designed to magnify distant objects. As in a microscope, the main parts of a telescope are its objective and eyepiece. Since the objects viewed through telescopes are very far away — essentially infinitely far — the light from the object is focused at the focal point of the objective. The telescope is designed so that this point is also the focal point of the eyepiece. With this information, it can be shown that the angular magnification of the telescope is given by

$$M = \frac{f_o}{f_e}$$

where f_o is the focal length of the objective and f_e is that of the eyepiece.

Practice Quiz

8. What effect is produced if the eyepiece of a telescope is replaced with one of shorter focal length?

(a) The image is enlarged.

(b) The image is diminished.

(c) The image becomes inverted.

(d) The focal length of the objective increases.

(e) The image is unaffected by changing the eyepiece.

9. The objective of a telescope has a focal length of 1.8 m. What is the focal length of an eyepiece that will produce a magnification of 500×?

(a) 2.0 mm **(b)** 278 m **(c)** 900 m **(d)** 3.6 mm **(e)** none of the above

27–6 Lens Aberrations

Only ideal lenses bring all light rays to a perfect focus. Real lenses tend to form blurred images by focusing different rays at different points. The inability of lenses to focus rays precisely is called **aberration**. Among the more commonly encountered forms of aberration is **spherical aberration,** which is due to the shape of lenses ground to spherical sections. **Chromatic aberration** is also commonly seen and is due to the dispersion of light in the lens. In many optical systems, chromatic aberration is reduced by using combinations of lenses.

Reference Tools and Resources

I. Key Terms and Phrases

accommodation the adjustment of the shape of the lens in a human eye to help focus objects

near point the closest distance to the eye at which an object can be placed and still be in focus

far point the greatest distance from the eye that an object can be, and still be in focus

aperture the opening through which light enters a camera

f-number a camera setting that determines the diameter of the aperture

shutter speed the amount of time that the shutter of a camera remains open

nearsightedness the ability to focus nearby objects but not distant ones

refractive power the inverse of the focal length of a lens

farsightedness the inability to focus nearby objects

angular magnification the ratio of an object's angular size seen with an optical device to its angular size when viewed at the near point with an unaided eye

objective the first lens through which light from the object comes into a telescope or a microscope

eyepiece a magnifying lens through which the observer views images in a telescope or a microscope

aberration the inability of real lenses to sharply focus light rays

II. Important Equations

Name/Topic	Equation	Explanation
cameras	$f\text{-number} = \frac{f}{D}$	The *f*-number setting of a camera
corrective optics	$\text{refractive power} = \frac{1}{f}$	The refractive power of a lens
magnifying glass	$M = \frac{\theta'}{\theta} = \frac{N}{f}$	The angular magnification of a magnifying glass for a relaxed eye
compound microscope	$M = -\frac{d_i N}{f_o f_e}$	The angular magnification of a compound microscope
telescope	$M = \frac{f_o}{f_e}$	The angular magnification of a refracting telescope

III. Know Your Units

Quantity	Dimension	SI Unit
refractive power	$[L^{-1}]$	diopter
angular magnification (M)	dimensionless	—

Puzzle

PHISHICS

A person with good vision finds that she cannot focus on anything underwater. However, plastic goggles with "lenses" that are *flat* plastic disks allow her to see fish clearly. Explain how this can be. (Hint: This has nothing to do with salt or chlorine in the water.) Answer this question in words, not equations, briefly explaining how you obtained your answer.

Answers to Selected Conceptual Questions and Exercises

Conceptual Questions

2. No. The lens will still show a complete image, though you may have to move your head more from side to side to see it all.

4. The reason things look blurry underwater is that there is much less refraction of light when it passes from water to your cornea than when it passes from air to your cornea. Therefore, your eyes simply aren't converging light enough when they are in water. Since farsightedness is caused when your eyes don't converge light as much as they should (see Figure 27-11), this can be considered as an extreme case of farsightedness.

8. No. Chromatic aberration occurs in lenses because light of different frequency refracts by different amounts. In the case of a mirror, however, all light – regardless of its frequency – obeys the same simple law of reflection; namely, that the angle of reflection is equal to the angle of incidence. Since light of all colors is bent in the same way by a mirror, there is no chromatic aberration.

Conceptual Exercises

4. If a person is nearsighted, the eye converges (bends) light too much to bring it to a proper focus on distant objects (see Figure 27-6). To reduce the amount of convergence, the intracorneal ring should decrease the cornea's curvature – that is, it should make the cornea flatter.

8. The instrument is a telescope. In general, the length of a telescope is roughly equal to the sum of the focal lengths of its objective and eyepiece, as we see in Figure 27-16.

12. As an object moves closer to the front of an octopus eye, the image it forms moves farther behind the eye. The situation is similar to that in Active Example 27-1 and Figure 26-35 (a). To keep the image on the retina, therefore, it is necessary to move the lens itself farther from the retina.

Solutions to Selected End-of-Chapter Problems

7. **Picture the Problem**: The aperture settings, or *f*-number of a camera lens, is the ratio of the focal length to the aperture diameter.

Strategy: Divide the focal length by the *f*-number to calculate the aperture diameter for each lens.

Solution: 1. (a) Since the diameter is inversely proportional to the *f*-number, the smallest *f*-number gives the largest diameter. So, $\boxed{2.8}$ gives the largest diameter.

2. (b) Calculate the diameter for *f*/2.8: $D = \dfrac{f}{f\text{-number}} = \dfrac{55\text{ mm}}{2.8} = \boxed{20\text{ mm}}$

3. Calculate the diameter for *f*/4: $D = \dfrac{55\text{ mm}}{4} = \boxed{14\text{ mm}}$

4. Calculate the diameter for *f*/8: $D = \dfrac{55\text{ mm}}{8} = \boxed{6.9\text{ mm}}$

5. Calculate the diameter for *f*/11: $D = \dfrac{55\text{ mm}}{11} = \boxed{5.0\text{ mm}}$

6. Calculate the diameter for *f*/16: $D = \dfrac{55\text{ mm}}{16} = \boxed{3.4\text{ mm}}$

Insight: As the aperture settings increase, the diameters decrease proportionately.

23. Picture the Problem: The relaxed cormorant's eye focuses on distant objects. By squeezing the lens its eye is able to change the refractive power by 45 diopters in order to focus on objects while underwater.

Strategy: Calculate the refractive power of the relaxed eye from the inverse of the focal length. Add the forty-five diopters to the refractive power and take the inverse to calculate the new focal length.

Solution: 1. (a) Lens refraction is lessened because n_{water} is closer to n_{lens} than n_{air} is. The compensating change in the lens's refractive power must be an $\boxed{\text{increase}}$.

2. (b) Calculate the relaxed refractive power: $rp_{\text{relaxed}} = \dfrac{1}{f_{\text{relaxed}}} = \dfrac{1}{0.0042\text{ m}} = 238\text{ diopters}$

3. Add the change in refractive power and invert to calculate the close focal length: $f_{\text{close}} = \dfrac{1}{rf_{\text{close}}} = \dfrac{1}{(238+45)\text{diopters}} = \boxed{3.5\text{ mm}}$

Insight: The focal length of the cormorant's eye ranges between 3.5 mm and 4.2 mm.

41. Picture the Problem: The refractive power of the distant-vision part of the bifocals creates an image of distant objects within the wearer's far point. The refractive power of the close-vision part of the bifocals creates an image of close objects outside the wearer's near point.

Strategy: Calculate the person's far point by adding the eye-to-lens distance to the negative of the image distance in Equation 26-16, where the object distance is infinity and the refractive power is that of the distant viewing. To calculate the near point, add the eye-to-lens distance to the negative of the image distance, where the object distance is the corrected near point (25.0 cm) minus the eye-to-lens distance and the refractive power is that of near viewing.

Solution: 1. (a) Calculate the image distance for distant viewing:

$$d_i = \left(\frac{1}{f} - \frac{1}{d_o}\right)^{-1} = \left(\frac{1}{f} - \frac{1}{\infty}\right)^{-1} = \left(-0.425 \text{ m}^{-1}\right)^{-1} = -2.35 \text{ m}$$

2. Add the eye-to-lens distance:

$$\text{far point} = -d_i + 0.0200 \text{ m} = -(-2.35 \text{ m}) + 0.0200 \text{ m} = \boxed{2.37 \text{ m}}$$

3. (b) Calculate the image distance for near viewing:

$$d_i = \left(\frac{1}{f} - \frac{1}{d_o}\right)^{-1} = \left(1.55 \text{ diopters} - \frac{1}{0.250 \text{ m} - 0.0200 \text{ m}}\right)^{-1} = -35.7 \text{ cm}$$

4. Add the eye-to-lens distance:

$$\text{near point} = -d_i + 0.0200 \text{ m} = -(-35.7 \text{ cm}) + 2.00 \text{ cm} = \boxed{37.7 \text{ cm}}$$

Insight: Without the glasses the person can only see clearly between 37.7 cm and 2.37 m. With the glasses the person can see clearly for all distances greater than 25.0 cm.

47. Picture the Problem: Two lenses with different focal lengths can each be used as simple magnifiers. The maximum magnification from either lens will occur when the virtual image is at the viewer's near point.

Strategy: Use Equation 27-5 to calculate the magnification from each lens.

Solution: 1. (a) Since a smaller focal length results in a larger magnification, the lens with focal length $\boxed{f_1}$ can produce the greater magnification.

2. (b) Calculate the magnification when f = 5.0 cm:

$$M_1 = 1 + \frac{N}{f} = 1 + \frac{25 \text{ cm}}{5.0 \text{ cm}} = \boxed{6.0}$$

3. Calculate the magnification when f = 13 cm:

$$M_2 = 1 + \frac{25 \text{ cm}}{13 \text{ cm}} = \boxed{2.9}$$

Insight: The smaller the focal length, the greater the possible magnification.

55. Picture the Problem: An objective lens of focal length 4.00 mm acting alone will produce a magnification of – 40.0. Adding an eyepiece to this objective lens will produce a microscope with a total magnification of 125x.

Strategy: Use Equation 26-18 to write the objective lens' image distance in terms of the lateral magnification and the object distance. Insert the image distance into Equation 26-16 and solve for the object distance. Finally, solve Equation 27-6 for the focal length of the eyepiece.

Solution: 1. (a) Write the image distance in terms of the magnification: $m = -d_{\text{i}}/d_{\text{o}}$

$d_{\text{i}} = -md_{\text{o}}$

2. Solve Equation 26-16 for the object distance:

$$\frac{1}{f_{\text{obj}}} = \frac{1}{d_{\text{o}}} + \frac{1}{d_{\text{i}}} = \frac{1}{d_{\text{o}}} - \frac{1}{md_{\text{o}}} = \frac{1}{d_{\text{o}}}\left(1 - \frac{1}{m}\right)$$

$$d_{\text{o}} = f_{\text{obj}}\left(1 - \frac{1}{m}\right) = 4.00\text{ mm}\left(1 - \frac{1}{-40.0}\right) = \boxed{4.10\text{ mm}}$$

3. (b) Use Eq. 27-6 to write the total magnification: $M_{\text{total}} = m_{\text{o}}M_{\text{e}} = m_{\text{o}}\dfrac{N}{f_{\text{e}}}$

4. Solve for the eyepiece focal length:

$$f_{\text{e}} = \frac{m_{\text{o}}N}{M_{\text{total}}} = \frac{-40.0(25\text{ cm})}{125} = \boxed{-8.0\text{ cm}}$$

Insight: To double the magnification, without changing the objective lens, would require an eyepiece with a focal length of –4.0 cm.

71. Picture the Problem: The figure shows a telescope with total length 275 mm and objective focal length 257 mm.

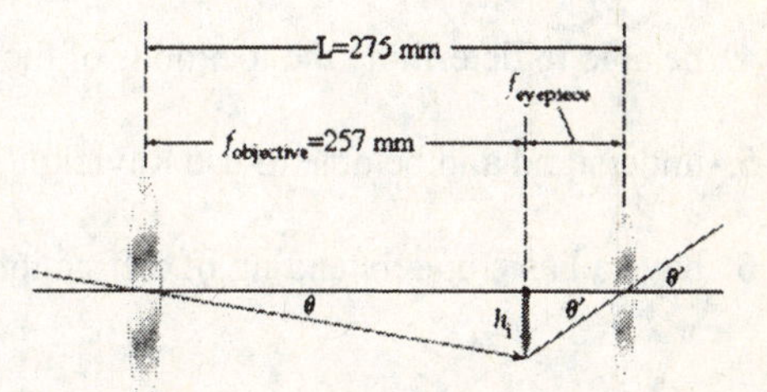

Strategy: Subtract the objective focal length from the total length to calculate the focal length of the eyepiece. Then use Equation 27-7 to calculate the angular magnification.

Solution: 1. (a) Calculate the focal length of the eyepiece:

$$f_{\text{e}} = L - f_{\text{obj}} = 275\text{ mm} - 257\text{ mm} = \boxed{18\text{ mm}}$$

2. (b) Use Equation 27-7 to calculate the magnification:

$$M_{\text{total}} = \frac{f_{\text{obj}}}{f_{\text{e}}} = \frac{257\text{ mm}}{18\text{ mm}} = \boxed{14}$$

Insight: To obtain a magnification of 20x with the same telescope length, the objective focal length would need to be increased to 262 mm.

Answers to Practice Quiz

1. (d) **2.** (a) **3.** (e) **4.** (b) **5.** (c) **6.** (d) **7.** (b) **8.** (a) **9.** (d)

CHAPTER 28
PHYSICAL OPTICS: INTERFERENCE and DIFFRACTION

Chapter Objectives

After studying this chapter, you should

1. understand Young's two-slit experiment in terms of wave interference for light.
2. be able to determine the locations of bright and dark fringes in the two-slit experiment.
3. be able to determine the conditions for constructive and destructive interference of reflected waves in air wedges and thin films.
4. be able to determine the locations of the dark fringes in single-slit diffraction.
5. understand and be able to use Rayleigh's criterion for the resolution of objects.
6. have a basic understanding of diffraction gratings.

Warm-Ups

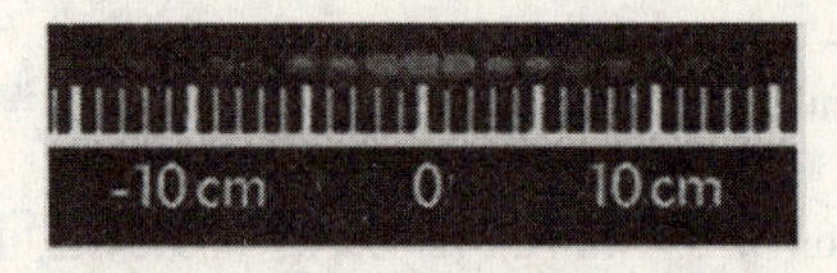

1. The accompanying picture shows light from a narrow laser beam as it appears on a screen after passing through two narrow slits. Examine the pattern and answer the following questions. What is the distance between two successive minima due to interference between the slits? What is the distance between two successive minima due to diffraction at the slits themselves? What would change if the separation between the slits was reduced? What would change if each slit was made wider?
2. The picture on the right shows what happens to a narrow laser beam after it passes through a small, round pinhole. The same happens to light passing through your pupil. In your textbook, find the criterion for resolving two light sources. Use the criterion to estimate at what minimum distance your eye is able to resolve the two headlights of an approaching car. (Estimate the diameter of your pupil to be 2 mm.) (Hint: Look at Example 28–6 in the text.)

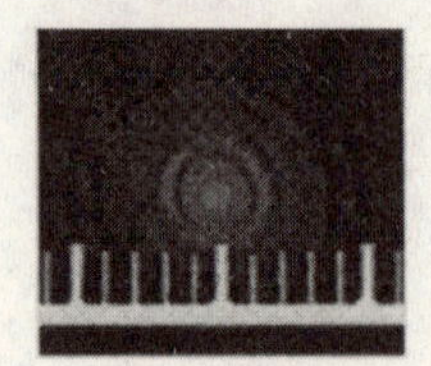

3. Why does a soap bubble reflect virtually no light just before it bursts?
4. Rock specimen slices are often 30 micrometers thick. Approximately how many wavelengths of visible light does that represent? What additional information would you need to be able to answer this question exactly?

Chapter Review

With geometrical optics, most of the situations we considered were such that the wave properties of light were of little direct consequence. In this chapter, we discuss aspects of the behavior of light that can only be understood in terms of its wave nature. This branch of optics is sometimes called *physical optics*.

28–1 Superposition and Interference

One of the key signatures of wave behavior is that of **superposition**, also called **interference**. This wave behavior for light is similar to what it is for sound waves, as discussed in Section 14–7. With sound, the superposition of waves occurs when different waves coexist in a region such that the net displacement at a point is the sum of the displacements of the individual waves. For light, the electric and magnetic fields play the role of the displacement. When the net fields resulting from the combination of waves have larger magnitudes than the fields from the individual waves, we call this **constructive interference**; when the combination results in fields of reduced magnitudes, we call this **destructive interference**.

Interference effects are noticeable when the different light waves are of the same frequency, or **monochromatic**, and have a constant phase relationship, or are **coherent**. When the phase difference between waves is 0°, or some multiple of 360° (corresponding to path differences that are multiples of the wavelength) the waves are *in phase*. When the phase difference is 180°, or some odd multiple of it (corresponding to path differences of odd multiples of a half wavelength) the waves are said to be *completely out of phase*. When the phase relationship between the different waves varies randomly, the waves are said to be **incoherent**.

Practice Quiz

1. If two light waves have equal wavelengths, but one travels a quarter of a wavelength farther in distance, what is the phase difference between these waves?

   (a) 0° **(b)** 25° **(c)** 30° **(d)** 45° **(e)** 90°

28–2 Young's Two-Slit Experiment

Young's two-slit experiment is a classic experiment that demonstrates the interference properties of waves. In this experiment, monochromatic light is passed through a single slit to produce a small source of light. This light then shines on a setup containing two slits, which act as independent coherent sources. By Huygens' principle, we can treat each slit as a source of light spreading out in all forward directions. The light from these two slits then shines on a screen. On the screen, there is a pattern of bright and dark *fringes* that are the result of constructive and destructive interference between the waves from the two slits.

In a typical setup, the distance from the slits to the screen is much larger than the separation between the slits. This assumption simplifies the analysis. There will be constructive interference on the screen, where there are bright fringes, when the path difference, $\Delta\ell$, between the light waves from the two slits equals an integral multiple of the wavelength, λ. Under these assumptions, this is equivalent to the condition

$$\sin\theta = m\frac{\lambda}{d}, \quad m = 0, \pm1, \pm2, \ldots \qquad \text{[bright fringes]}$$

where θ is the angle from either slit to the relevant point on the screen (which are approximately equal under the current assumptions), and d is the separation between the slits. The dark fringes result from destructive interference that occurs for path differences that are odd multiples of $\lambda/2$, which leads to the condition

$$\sin\theta = \left(m - \tfrac{1}{2}\right)\frac{\lambda}{d}, \quad m = 0, \pm1, \pm2, \ldots \qquad \text{[dark fringes]}$$

At the center of the interference pattern on the screen is a bright fringe. The position of a given fringe is determined by its linear distance on the screen from the center of this central bright fringe. This linear distance is given by

$$y = L\tan\theta$$

where L is the distance between the screen and the plane of the two slits, and θ is the angle measured from the spot exactly between the slits to the fringe being located on the screen; under the assumption that $L >> d$, this angle is approximately equal to the angle found from the previous two conditions for the bright and dark fringes.

Example 28–1 A Two-Slit Experiment Light of wavelength 455 nm is incident on a two-slit apparatus with a slit separation distance of 0.125 mm. What is the distance on a screen, 2.00 m away, between the first and fifth dark fringes?

Picture the Problem The diagram shows a two-slit setup and indicates the distance y between the first and fifth dark fringes.

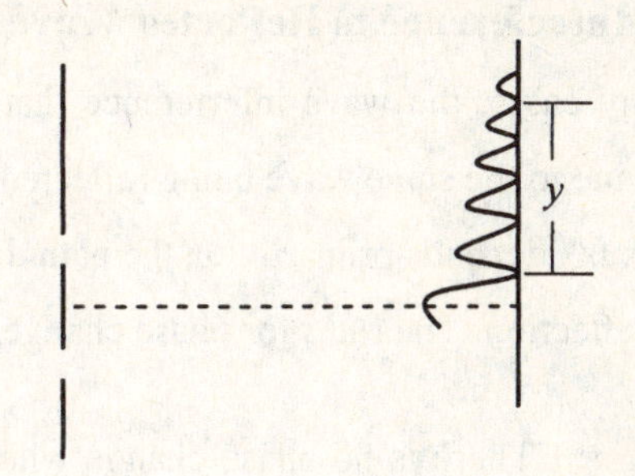

Strategy Because we know which dark fringes we are considering, we can determine their order numbers and their angles. We can then calculate the difference in their distances on the screen.

Solution

1. The first dark fringe has $m = +1$, so its angle is: $\sin\theta_1 = (1 - \frac{1}{2})\frac{\lambda}{d} \Rightarrow \theta_1 = \sin^{-1}\left(\frac{\lambda}{2d}\right) = 0.10428^\circ$

2. The fifth dark fringe has $m = +5$, so its angle is: $\sin\theta_5 = (5 - \frac{1}{2})\frac{\lambda}{d} \Rightarrow \theta_5 = \sin^{-1}\left(\frac{9\lambda}{2d}\right) = 0.93855^\circ$

3. The linear distance is determined by: $y = y_5 - y_1 = L[\tan\theta_5 - \tan\theta_1]$

4. Solving for the numerical result gives: $y = (2.00\text{ m})\left[\tan(0.93855^\circ) - \tan(0.10428^\circ)\right] = 0.0291\text{ m}$

Insight This problem also could have been done using $m = 0$ and $m = -4$.

Practice Quiz

2. Monochromatic light of wavelength λ incident on a two-slit apparatus produces an angular separation ϕ between bright fringes. If a light wave of wavelength $\lambda/2$ is used, which of the following statements is true about the angular separation between bright fringes?
   (a) It doubles to 2ϕ.
   (b) It increases to greater than ϕ, but not necessarily 2ϕ.
   (c) It is cut in half to $\phi/2$.
   (d) It decreases to less than ϕ, but not necessarily $\phi/2$.
   (e) None of the above.

3. Light of wavelength 650 nm is incident on a two-slit apparatus with slit separation of 0.15 mm. At what angle will you find the first dark fringe?
   (a) 0.15° **(b)** 65 **(c)** 0.12° **(d)** 0.0022° **(e)** none of the above

28–3 Interference in Reflected Waves

In many cases, the wave interference that we observe is not due to waves from different sources, but rather due to the same wave being reflected from different surfaces or different locations. The interference that results depends primarily on the path difference between the waves and phase changes that may occur upon reflection. The rules for phase changes upon reflection are the following:

* There is no phase change when light reflects from the boundary of a region with a lower index of refraction than that for the incident light.

* There is a 180° (half wavelength) phase change when light reflects from the boundary of a region with a higher index of refraction than that for the incident light.

Air Wedge

One example where the interference between reflected waves occurs is found with an air wedge, which results when two plates of glass touch at one end and have a small separation at the other end. This setup creates a wedge-shaped air pocket between the plates. For this case, we consider the dominant interference effect, which is between light reflected from the bottom surface of the top plate and light reflected from the top surface of the lower plate. The bottom of the upper plate is a glass-to-air boundary, so there is no phase change upon reflection ($n_{glass} > n_{air}$) for these rays. The top of the lower surface is an air-to-glass boundary, so there is a 180° phase change upon reflection from this surface. This phase change contributes a half wavelength to the *effective* path difference between the rays. The rest of the effective path difference comes from the fact that, for normal incidence, one of the rays travels an extra distance of $2d$, where d is the thickness of the air gap. Thus, the effective path difference is

$$\Delta\ell_{eff} = \frac{\lambda}{2} + 2d$$

If this effective path difference is an integral number of wavelengths, there will be constructive interference:

$$\frac{1}{2} + \frac{2d}{\lambda} = m, \quad m = 1,2,3,\ldots \qquad \text{[constructive interference]}$$

If the effective path difference equals an odd multiple of a half wavelength, there will be destructive interference:

$$\frac{1}{2} + \frac{2d}{\lambda} = m + \tfrac{1}{2}, \quad m = 0,1,2,\ldots \qquad \text{[destructive interference]}$$

Thin Films

The interference of light due to thin films is very commonly seen and widely used in optics; almost every lens in any optical system is coated with a film in order to take advantage of these effects. The conditions for constructive and destructive interference are found by similar reasoning as with an air wedge. For thin films, we consider the interference between a ray reflected at the top surface of the film, ray 1, and a ray reflected at the bottom surface of the film, ray 2.

The top surface of the film is an air-to-film boundary, so there is a 180° phase shift for ray 1. Whether there will be a phase shift at the bottom surface depends on whether the index of refraction of the film is less than or greater than that of the *substrate* on which the film sits. If there is a phase shift for ray 2, it effectively cancels out the phase shift for ray 1. If there is no phase shift for ray 2, then there is a contribution of a half wavelength to the effective path difference. Once we know the phase difference due to the phase shifts (either 0 or 180°), we must consider the contribution resulting from the actual path difference, since ray 2 travels an extra distance of approximately twice the thickness of the film, t (for nearly normal incidence).

For ray 2, we must keep in mind that in the film the wavelength of light will be different from that in air (or vacuum) according to

$$\lambda_{film} = \frac{\lambda_{vac}}{n_{film}}$$

Therefore, the conditions for constructive or destructive interference occur when the distance that ray 2 travels in the film, $2t$, is an integral multiple of λ_{film} or an odd multiple of $\lambda_{film}/2$. The actual conditions are determined on a case-by-case basis.

Example 28–2 Thin-Film Coatings You need to reduce the glare from a lens in an optical system. The index of refraction of the lens is 1.50. To reduce the glare from visible light you decide to coat the lens with a substance whose index of refraction is 1.35. Focusing on the wavelength of light at the middle of the visible part of the electromagnetic spectrum, 550 nm, with what minimum thickness should you coat the lens?

Picture the Problem The sketch shows the incident and reflected rays from the top and bottom boundaries of the thin film.

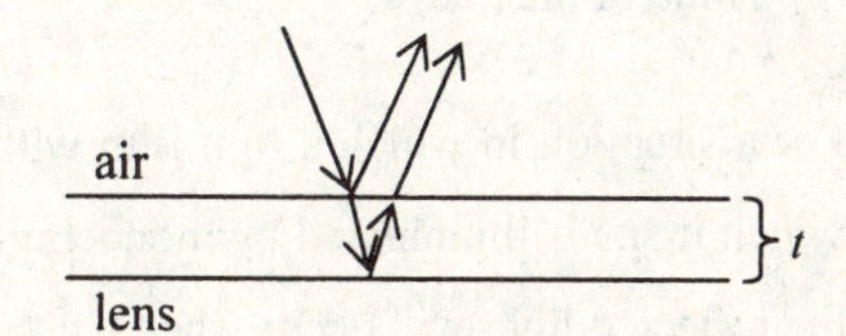

Strategy We want destructive interference of the

reflected waves. First, we consider the possible phase shifts upon reflection and then the needed path difference.

Solution

1. At the top boundary $n_{film} > n_{air}$, so: There is a 180° phase shift.

2. At the bottom boundary $n_{film} < n_{lens}$ so: There is a 180° phase shift.

3. The condition for the minimum thickness due to the path difference is: $2t_{\min} = \frac{\lambda_{film}}{2} = \frac{\lambda_{vac}}{2n_{film}}$

4. Solving for the thickness gives: $t_{\min} = \frac{\lambda_{vac}}{4n_{film}} = \frac{550\text{ nm}}{4(1.35)} = 102\text{ nm}$

Insight There are many other values of t that would produce destructive interference. See if you can find one more.

Practice Quiz

4. A thin film with an index of refraction of 1.66 covers a region of water ($n = 1.33$), with air being above the film. Which of the following statements is true about phase shifts upon reflection?
   (a) There is a 180° phase shift for light reflected off the top surface of the film and for light reflecting off the film-water boundary.
   (b) There is no phase shift for light reflected neither off the top surface of the film nor off the film-water boundary.
   (c) There is no phase shift for light reflected off the top surface of the film, but there is a 180° phase shift for light reflected off the film-water boundary.
   (d) There is a 180° phase shift for light reflected off the top surface of the film, but there is no phase shift for light reflected off the film-water boundary.
   (e) None of the above.

5. For a situation in which a thin film with an index of refraction greater than that of the substance on which it sits is illuminated by monochromatic light incident from the air, constructive interference for the reflected light will occur when the thickness of the film equals
   (a) $\lambda_{film}/4$. **(b)** $\lambda_{film}/2$. **(c)** $\lambda_{air}/4$. **(d)** $\lambda_{air}/2$. **(e)** $3\lambda_{air}/4$.

6. For a situation in which a thin film with an index of refraction less than that of the substance on which it sits is illuminated by monochromatic light incident from the air, destructive interference for the reflected light will occur when the thickness of the film equals

   (a) $\lambda_{film}/4$. **(b)** $\lambda_{film}/2$. **(c)** λ_{film}. **(d)** $\lambda_{air}/4$. **(e)** $\lambda_{air}/2$.

28–4 Diffraction

Besides interference, another characteristic of wave behavior is that waves bend when they pass by barriers or through openings; this phenomenon is known as **diffraction**. In this section, we consider the diffraction of monochromatic light as it passes through a single small slit or opening of width W. If this diffracted light shines on a distant screen, we see an interference pattern on the screen. This pattern can be understood in terms of Huygens's principle in which every point within the slit can be treated as a separate source of light waves in all forward directions. The interference pattern is then a result of the superposition of all these waves. Taking $\theta = 0$ to be the direction directly along a line passing through the slit and perpendicular to the distant screen, the angular locations of the dark fringes of the interference pattern are determined by

$$\sin\theta = m\frac{\lambda}{W}, \quad m = \pm 1, \pm 2, \pm 3, \ldots \qquad \text{[dark fringes]}$$

The numbers m take both positive and negative values to account for the fact that the diffraction pattern is symmetric, so that there are dark fringes on both sides of a central bright fringe.

Exercise 28–3 Single-Slit Diffraction Light of wavelength 675 nm is incident on a single slit of width 3.15 μm. What is the angular width of the bright fringe that is closest to the central maximum?

Solution

The width, $\Delta\theta$, of the first bright fringe (or first-order bright fringe) can be determined by the locations of the dark fringes on either side of it. The first- and second-order dark fringes correspond to $m = 1$ and $m = 2$, and they are located at angles

$$\theta_1 = \sin^{-1}\left(\frac{\lambda}{W}\right) \quad \text{and} \quad \theta_2 = \sin^{-1}\left(\frac{2\lambda}{W}\right)$$

Therefore,

$$\Delta\theta = \theta_2 - \theta_1 = \sin^{-1}\left(\frac{2\lambda}{W}\right) - \sin^{-1}\left(\frac{\lambda}{W}\right) = \sin^{-1}\left(\frac{2(675\times10^{-9}\text{ m})}{3.15\times10^{-6}\text{ m}}\right) - \sin^{-1}\left(\frac{675\times10^{-9}\text{ m}}{3.15\times10^{-6}\text{ m}}\right) = 13.0^\circ$$

Practice Quiz

7. The effects of diffraction in a single slit will become more pronounced if
   (a) the width of the slit decreases.
   (b) the width of the slit increases.
   (c) the wavelength of the light is decreased.
   (d) the width of the slit has no effect.
   (e) None of the above.

28–5 Resolution

Light enters optical instruments, such as the human eye, cameras, and telescopes, through apertures (such as the pupil of a human eye). This light, therefore, experiences diffraction, and this diffraction limits our ability to resolve images. The ability to visually distinguish closely spaced objects is called **resolution.** For a circular aperture of diameter D, the condition for the first dark fringe is given by

$$\sin\theta = 1.22\frac{\lambda}{D}$$

Rayleigh's criterion for the ability to resolve two objects is that the angular separation of their central bright spots must be greater than that of their first dark fringes. Because for small angles, $\sin\theta \approx \theta$, we write this criterion for limiting resolution as

$$\theta_{\min} = 1.22\frac{\lambda}{D}$$

where $\theta_{\min}$ is in radians.

Practice Quiz

8. The ability to resolve two objects will improve if
   (a) the diameter of the aperture decreases.
   (b) the diameter of the aperture increases.
   (c) the diameter of the aperture equals the wavelength of the light from the objects.
   (d) the wavelength of the light is longer.
   (e) None of the above.

28–6 Diffraction Gratings

The diffraction of light plays an important role in Young's double-slit experiment. Increasing the number of slits creates an interference pattern (or diffraction pattern) that contains sharper and more widely spaced principal maxima. A system with a large number of slits for the purpose of analyzing light sources is called a **diffraction grating**. The angles at which these principal maxima occur for a diffraction grating is given by

$$\sin\theta = m\frac{\lambda}{d}, \quad m = 0, \pm 1, \pm 2, \ldots$$

where d is the separation between adjacent slits of the grating, and λ is the wavelength of the incident monochromatic light. Typically, diffraction gratings are described in terms of the number of lines (slits) per unit length, N. From this value, the slit separation can be found as $d = 1/N$.

Example 28–4 A Diffraction Grating Your company is under contract to manufacture a diffraction grating for which the first-order ($m = 1$) maximum, for light of wavelength 475 nm, occurs at an angle of 25° away from the central maximum. How many lines per centimeter should your grating have?

Picture the Problem The sketch shows the diffraction grating and the first-order principal maximum.

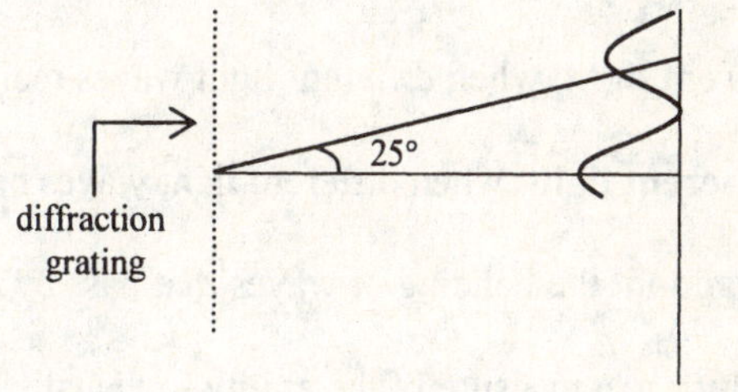

25°

diffraction grating

Strategy We can use the expression for the principal maxima to determine the slit spacing, from which we can calculate N.

Solution

1. The slit spacing is given by: $\sin\theta = m\frac{\lambda}{d} \Rightarrow d = \frac{m\lambda}{\sin\theta}$

2. The number of lines per unit length is: $N = \frac{1}{d} \Rightarrow N = \frac{\sin\theta}{m\lambda}$

3. Obtain the numerical result: $N = \frac{\sin\theta}{m\lambda} = \frac{\sin(25^\circ)}{475\times10^{-7}\text{ cm}} = 8900\text{ lines/cm}$

Insight The main thing to be careful about here is the units in which N is to be expressed. We need to make sure that if we want lines/cm, we don't actually calculate lines/m.

Practice Quiz

9. A diffraction grating has 3310 lines/cm. What is the distance between its slits?

   (a) 3.02 mm **(b)** 3.02 cm **(c)** 0.302 cm **(d)** 3.02 m **(e)** None of the above.

10. Light of wavelength 650 nm is incident on a diffraction grating with 6666 lines/m. At what angle will you find a first-order principal maximum?

(a) 0.15° **(b)** 26° **(c)** 5.6° **(d)** 0. 25° **(e)** None of the above.

Reference Tools and Resources

I. Key Terms and Phrases

superposition/interference the combination of waves such that the net "displacement" at a point is the sum of the displacements of the individual waves

constructive interference when waves superimpose such that the net displacements are increased

destructive interference when waves superimpose such that the net displacements are diminished

monochromatic light light of a single frequency (or wavelength)

coherent light when different light waves maintain a constant phase relationship

incoherent light when different light waves have a random phase relationship

diffraction the bending of waves that pass by obstacles or through openings

resolution a measure of the ability to visually distinguish closely spaced objects

diffraction grating a system with a large number of slits for the purpose of analyzing light sources

II. Important Equations

Name/Topic	Equation	Explanation
Young's two-slit experiment	$\sin\theta = m\frac{\lambda}{d}, \quad m = 0, \pm1, \pm2, \ldots$	The condition for bright fringes
	$\sin\theta = (m - \frac{1}{2})\frac{\lambda}{d}, \quad m = 0, \pm1, \pm2, \ldots$	The condition for dark fringes
single-slit diffraction	$\sin\theta = m\frac{\lambda}{W}, \quad m = \pm1, \pm2, \pm3, \ldots$	The condition for dark fringes
resolution	$\theta_{\min} = 1.22\frac{\lambda}{D}$	Rayleigh's criterion for the condition when two objects can be barely resolved
diffraction gratings	$\sin\theta = m\frac{\lambda}{d}, \quad m = 0, \pm1, \pm2, \ldots$	The condition for the principal maxima of a diffraction grating

Puzzle

ONE POLARIZER, TWO POLARIZER

Two crossed polarizing filters (oriented with their axes at right angles to each other) almost completely annihilate a light beam (initially unpolarized). If a third polarizer is placed between the two, the light intensity usually increases. Why is that? How would you orient the middle polarizer to maximize the intensity of transmitted light? What percentage of the original unpolarized beam would pass through such a setup?

Answers to Selected Conceptual Questions and Exercises

Conceptual Questions

2. If the slit spacing, *d*, were less than the wavelength, λ, the condition for a bright fringe (Equation 28-1) could be satisfied only for the central bright fringe ($m = 0$). For nonzero values of *m* there are no solutions, because sin *q* cannot be greater than one. In addition, Equation 28-2 shows that if *d* is greater than $\lambda/2$, though still less than λ, there will be only one dark fringe on either side of the central bright fringe. If *d* is less than $\lambda/2$, no dark fringes will be observed.

6. Submerging the two-slit experiment in water would reduce the wavelength of the light from λ to λ/n, where $n = 1.33$ is the index of refraction of water. Therefore, the angles to all the bright fringes would be reduced, as we can see from Equation 28-1. It follows that the two-slit pattern of bright fringes would be more tightly spaced in this case.

12. In an iridescent object, the color one sees is determined by constructive and destructive interference. The conditions for interference, however, depend on path length, and path length depends on the angle from which one views the system. This is analogous to viewing a two-slit system from different angles and seeing alternating regions of constructive and destructive interference. A painted object, on the other hand, simply reflects light of a given color in all directions.

Conceptual Exercises

4. The physical path difference for the two rays is $2d$. In addition, each of the rays will have a phase change of half a wavelength if it reflects from a region with a larger index of refraction. **(a)** The physical path difference is $\lambda/2$, therefore one of the rays must undergo an additional phase change of half a wavelength to result in constructive interference. It follows, then, that constructive interference will be observed in case 3 and case 4. **(b)** In this case the physical path difference is λ.

Therefore, to obtain constructive interference, both rays must have the same phase change due to reflection. As a result, constructive interference occurs in cases 1 and 2.

6. Light reflected from the top of the film has a phase change of 180°; light reflected from the film-water surface also has a phase change of 180°, since the film's index of refraction is less than that of water. It follows that the film appears bright (constructive interference) where the film's thickness goes to zero.

10. Reducing the index of refraction would increase the wavelength of light within the eye. We know from Equation 28-14, however, that resolution decreases as the wavelength increases; that is, increasing λ. in Equation 28-14 implies that the required angle of separation, θ, must also increase. Therefore, the eye's resolution would decrease in this case.

Solutions to Selected End-of-Chapter Problems

11. Picture the Problem: The figure shows a car traveling at 17 m/s perpendicular to the line connecting two transmitter towers. When between the towers the car picks up a minimum signal.

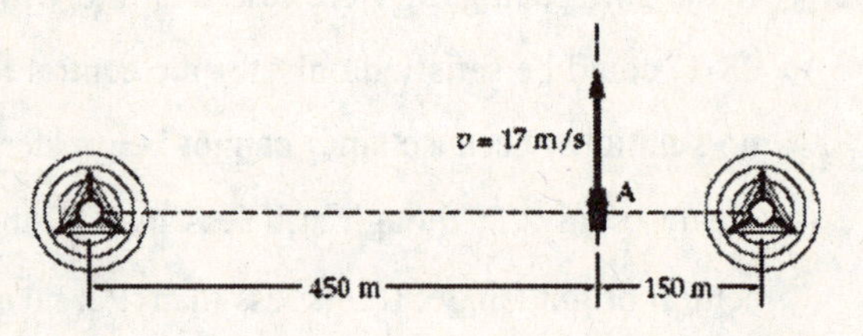

Strategy: Set a difference in distance between the car and each tower equal to a half wavelength for the longest wavelength that will produced destructive interference and equal to three half wavelengths for the next longest wavelength.

Solution: 1. (a) Calculate the difference in distance between the two towers: $\Delta\ell = 450\text{ m} - 150\text{ m} = 300\text{ m}$

2. Set the distance equal to a half wavelength: $\frac{1}{2}\lambda = 300\text{ m}$

$\lambda = 2(300\text{ m}) = \boxed{600\text{ m}}$

3. (b) Since the path difference will equal the wavelength, the waves will arrive in phase, producing a $\boxed{\text{maximum}}$.

4. (c) Set the distance equal to three half wavelengths: $\frac{3}{2}\lambda = 300\text{ m}$

$\lambda = \frac{2}{3}(300\text{ m}) = \boxed{200\text{ m}}$

Insight: In general the wavelengths that produced destructive interference are $\lambda_n = \dfrac{600\text{ m}}{(2n+1)}$ for n=0,1,2,3,....

15. Picture the Problem: The figure shows light passing through a two slits separated by 48.0 x 10^{-5} m. The second order maximum is 0.0990° from the central maximum.

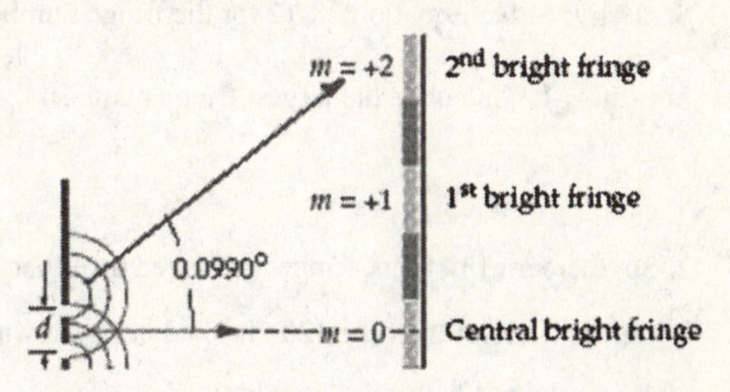

Strategy: Solve Equation 28-1 for the wavelength with $m = 2$ for the second bright fringe.

Solution: 1. (a) Calculate the wavelength: $m\lambda = d\sin\theta$

$$\lambda = \frac{d\sin\theta}{m} = \frac{(48.0\times10^{-5}\text{ m})\sin 0.0990°}{2} = \boxed{415\text{ nm}}$$

2. (b) Since the wavelength is directly proportional to the separation, the wavelength will $\boxed{\text{increase}}$ if the separation is increased.

3. (c) Calculate the wavelength for the larger slit separation: $\lambda = \dfrac{(68.0\times10^{-5}\text{ m})\sin 0.0990°}{2} = \boxed{587\text{ nm}}$

Insight: In general, the wavelength can be written as $\lambda = 8.64\times10^{-4}d$.

33. Picture the Problem: The image shows light at normal incidence reflecting off a thin layer of magnesium fluoride (n=1.38) attached to the surface of flint glass (n=1.61) The film suppresses the reflection of 565 nm light.

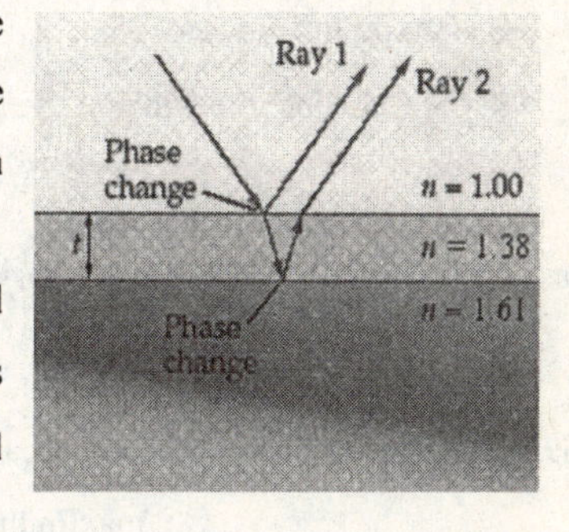

Strategy: Set the phase change for reflection at the air-film interface equal to $1/2$, and the phase change for reflection at the film-glass interface as $2t/\lambda_n + 1/2$. Subtract these two phase differences and set the result equal to $1/2$ for the minimum thickness that gives destructive interference. Solve the resulting equation for the thickness of the coating.

Solution: 1. (a) Set the phase differences equal to one-half: $\left(\dfrac{2t}{\lambda_n} + \dfrac{1}{2}\right) - \dfrac{1}{2} = \dfrac{1}{2}$

2. Solve for the coating thickness: $t = \dfrac{\lambda_n}{4} = \dfrac{\lambda}{4n} = \dfrac{565\text{ nm}}{4(1.38)} = \boxed{102\text{ nm}}$

3. (b) Higher frequency corresponds to shorter wavelength. A shorter wavelength would require a $\boxed{\text{thinner}}$ coating.

Insight: To block a frequency of 6.8×10^{14} Hz, which corresponds to a wavelength of 440 nm, a coating 79.7 nm is needed.

43. Picture the Problem: Green light (λ=553 nm) is incident upon a 8.00 μm slit.

Strategy: Solve Equation 28-12 for the fringe number closest to $\theta = 90°$.

Solution: 1. Calculate the largest fringe number:

$$\sin 90° = 1 = m\frac{\lambda}{W}$$

$$m = \frac{W}{\lambda} = \frac{8000 \text{ nm}}{553 \text{ nm}} = 14.47 \approx 14$$

2. So, there are $\boxed{14}$ dark fringes produced on either side of the central maximum.

Insight: If violet light (λ=420 nm) had been shown on the slit, there would have been 19 dark fringes on each side of the central maximum.

51. Picture the Problem: A lens has an aperture of 130 mm and focal length 640 mm. 540 nm light passes through the lens.

Strategy: Calculate the angle from the central maximum to the first minimum using Equation 28-14. Double this angle to calculate the angular width of the central maximum. Then set the width of the central maximum at the focal length equal to twice the distance given by Equation 28-3, where L is the focal length.

Solution: 1. Calculate the angle to the first minimum:

$$\sin\theta_{\text{min}} \approx \theta_{\text{min}} = 1.22\frac{\lambda}{D} = 1.22\frac{540 \text{ nm}}{0.130 \text{ m}} = 5.07\times10^{-6} \text{ rad}$$

2. Calculate the angular width of the central maximum:

$$\theta = 2\theta_{\text{min}} = 2\left(5.07\times10^{-6} \text{ rad}\right) = \boxed{1.0\times10^{-5} \text{ rad}}$$

3. (b) Calculate the width at the focal length.

$$2y = 2L\tan\theta_{\text{min}} \approx 2L\theta_{\text{min}} = 2\left(640\times10^{-3}\text{ m}\right)\left(5.07\times10^{-6}\text{rad}\right) = \boxed{6.5\ \mu\text{m}}$$

Insight: Diffraction, caused by the finite size of the lens, cause light the spread out so that it does not all focus at the focal point.

59. Picture the Problem: Light passing through a grating with 560 lines per centimeter has a second order maximum at 3.1°.

Strategy: Solve Equation 28-16 for the wavelength of light, where the slit separation, d, is the inverse of the number of lines per centimeter, N.

Solution: 1. Calculate the wavelength:

$$\sin\theta = \frac{m\lambda}{d} = m\lambda N$$

$$\lambda = \frac{\sin\theta}{mN} = \frac{\sin 3.1°}{2\left(560 \text{ cm}^{-1}\right)} = 4.8\times10^{-5}\text{cm} = \boxed{480 \text{ nm}}$$

2. (b) Since $\sin\theta$ is proportional to the number of lines per centimeter of the grating, a larger number of lines per centimeter results in a larger second-order maximum angle. So, the angle is $\boxed{\text{greater than}}$ 3.1°.

Insight: If the grating ha 1560 lines per centimeter the second order maximum would be at 8.7°.

Answers to Practice Quiz

1. (e) **2.** (d) **3.** (c) **4.** (d) **5.** (a) **6.** (a) **7.** (a) **8.** (b) **9.** (e) **10.** (d)

CHAPTER 29
RELATIVITY

Chapter Objectives

After studying this chapter, you should

1. be able to calculate the time interval between events relative to the proper time.
2. be able to calculate the length of an object relative to the proper length.
3. be able to use the relativistic addition of velocities for one-dimensional motion.
4. be familiar with the expressions for relativistic momentum and energy.

Warm-Ups

1. Einstein's theory of special relativity predicts that the rate at which a moving clock runs will be slower than the rate of the same clock when stationary. Suppose you are shut in a rocket module which is moving through space at half the speed of light. Would you be able to observe the "slowing down" of the clock you brought with you from Earth?
2. Suppose you had a car engine that could convert the rest energy of ordinary water into kinetic energy of the car. Estimate the amount of water needed to power an average car for the lifetime of the car. (Use reasonable estimates for the number of useful years and for average yearly mileage.)
3. According to the theory of relativity, the momentum formula $p = mv$ has to be modified with the factor gamma. Which of the sketches of p vs. v/c best agrees with the relativistic momentum formula? Why?

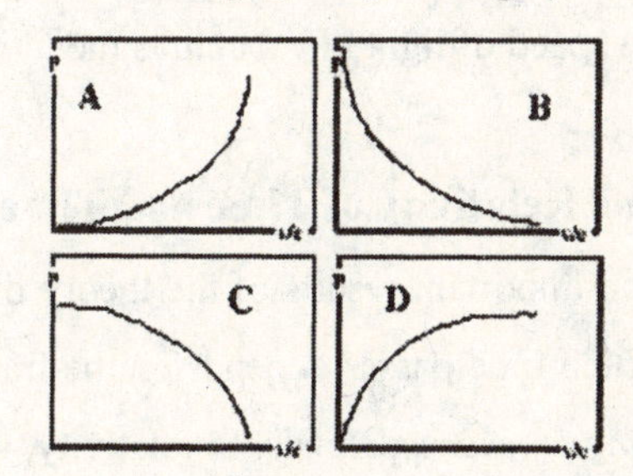

4. Estimate how fast you would have to run to have a relativistic mass of 1000 kg.

Chapter Review

It turns out that Newton's laws, despite their incredible success at describing physical behavior, are fundamentally incorrect. One of the corrections is contained in Einstein's theory of relativity. This theory is divided into two categories known as *special relativity* and *general relativity*. Special relativity is the main topic of this chapter, while general relativity is only briefly discussed. (The other major correction to Newton's laws is quantum physics, which is the topic of the next chapter.)

29–1 The Postulates of Special Relativity

The theory of special relativity is based on the following two postulates:

Equivalence of Physical Laws

The laws of physics are the same in all inertial frames of reference.

Constancy of the Speed of Light

The speed of light in vacuum, $c = 3.00 \times 10^8$ m/s, is the same in all inertial frames of reference, independent of the motion of the source or the receiver.

Recall that an inertial reference frame is one in which Newton's laws of motion are obeyed. Inertial frames of reference move with constant velocity relative to each other. The first postulate basically says that all the laws of physics are independent of the inertial reference frame in which they are being investigated.

The second postulate expresses the astounding fact that the speed of light in vacuum will have the same value, c, in all inertial frames of reference. As we will see, the consequences of this postulate are that the behavior of space and time is very different from what our everyday experience suggests. The reason for this difference is that for the speeds at which we move around in everyday life these unexpected effects are too small to be commonly noticed. Another important result from these postulates is that the speed of light in vacuum is the ultimate speed at which any object can move.

29–2 The Relativity of Time and Time Dilation

One of the important results of the theory of relativity is our knowledge that time is relative; that is, the rate at which time passes depends on the frame of reference. The basic result, known as **time dilation**, is that *moving clocks run slowly.* In relativity, we think of space and time in terms of **events**, where an event is just a particular point in space at a particular time. (If you prefer, you can think of something happening at that point and at that time, but it isn't necessary.) The amount of time between two events that are at the same point in space is called the **proper time**, Δt_0. The expression for how the time interval between events as measured in an arbitrary inertial frame, Δt, relates to the proper time is

$$\Delta t = \frac{\Delta t_0}{\sqrt{1 - v^2/c^2}}$$

where v is the relative velocity between the two inertial frames. For objects moving at everyday speeds, the smallness of the time dilation effect is such that most calculators will just give $\Delta t = \Delta t_0$. To handle these cases, it is better to use the approximate equation

$$\Delta t = \Delta t_0 \left(1 + \frac{1}{2}\frac{v^2}{c^2} \cdots\right)$$

An additional consequence of the relativity of time is that simultaneity is relative; that is, two events that occur at the same time in one frame of reference will not generally be simultaneous in another frame of reference.

Example 29–1 A Moving Clock If exactly 1.0 hour has passed according to the clock on your wall, how much time will have passed according to someone moving past you at 93% of the speed of light?

Picture the Problem The diagram shows your clock and the observer moving by at speed v.

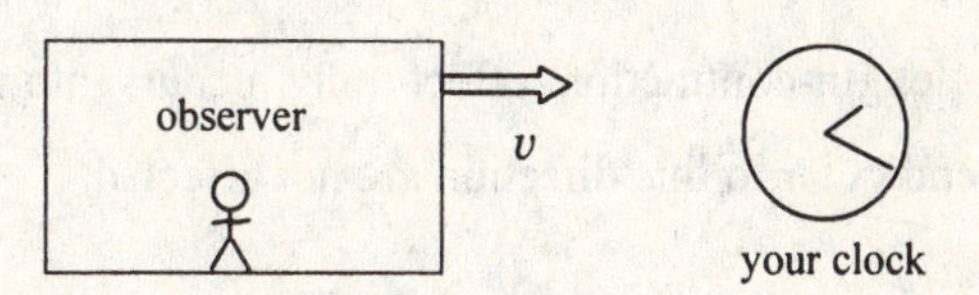

Strategy The time you measure on your clock is the proper time, Δt_0.

Solution

1. According to the time dilation equation:

$$\Delta t = \frac{\Delta t_0}{\sqrt{1 - v^2/c^2}} = \frac{\Delta t_0}{\sqrt{1 - (0.93c)^2/c^2}}$$

$$= \frac{1.0\text{ h}}{\sqrt{1 - (0.93)^2}} = 2.7\text{ h}$$

Insight Notice that to the observer, it is *your* clock that is moving, so she sees your clock as being slow because 2.7 h has passed in her reference frame.

Practice Quiz

1. The statement that moving clocks run slowly implies that
   (a) people moving relative to you age more slowly than you do.
   (b) people moving relative to you age more rapidly than you do.
   (c) people moving relative to you age at the same rate as you do.
   (d) This statement doesn't apply to the aging of people.
   (e) None of the above.

2. For what relative velocity would moving clocks run at half the rate of stationary clocks?

(a) $0.50c$ **(b)** $0.87c$ **(c)** $0.25c$ **(d)** $1.7c$ **(e)** $0.75c$

29–3 The Relativity of Length and Length Contraction

As with time, space is also relative. By this we mean that spatial distance, or length, depends on the frame of reference of the observer. This phenomenon is called **length contraction.** It has this name because the length of a moving object is shorter, along the direction of relative motion, than it would be if it were stationary. The length of an object as measured in a frame of reference at rest with respect to the object (the *proper frame*) is called its **proper length,** L_0. The length of this object as measured in an inertial frame of reference moving with speed v relative to the proper frame is given by

$$L = L_0\sqrt{1-\frac{v^2}{c^2}}$$

This length-contraction effect only occurs along the direction of relative motion; lengths that are perpendicular to this direction are not affected.

Example 29–2 A Moving Rod Suppose, through sophisticated techniques, you measure the length of a rod that is moving by you at 93.0% of the speed of light to be 3.30 m. What length would you measure for this same rod if it were in your frame of reference?

Picture the Problem The diagram shows the moving rod in its reference frame and you.

rod

v

you

Strategy If the rod were in your reference frame, you would measure the proper length, L_0.

Solution

1. Solve the length contraction equation for the proper length:

$$L = L_0\sqrt{1-\frac{v^2}{c^2}} \Rightarrow L_0 = L\left(1-\frac{v^2}{c^2}\right)^{-1/2}$$

2. Obtain the numerical result:

$$L_0 = L\left(1-\frac{v^2}{c^2}\right)^{-1/2} = (3.30\text{ m})\left(1-[0.930]^2\right)^{-1/2} = 8.98\text{ m}$$

Insight You would measure the moving rod as shorter, so it makes sense that its proper length is longer.

Practice Quiz

3. The fact that the length of a moving object is shortened implies that

   (a) the proper length is always the shortest length you can measure.

   (b) the proper length is always the longest length you can measure.

   (c) you can't measure a moving length.

   (d) the length of a moving object only appears to be shorter, but it is not actually shorter.

   (e) None of the above.

4. If the proper length of a rod that is laid out along the x direction is measured to be 1.0 m, how long will it be as measured in a frame moving at $0.90c$ relative to the proper frame along the y direction?

   (a) 1.0 m **(b)** 2.3 m **(c)** 0.9 m **(d)** 1.8 m **(e)** 3.2 m

29–4 The Relativistic Addition of Velocities

As stated above, relativity tells us that no object can travel faster than the speed of light in vacuum. This implies that the rules for velocity addition that we used previously (see Section 3–6) must be corrected by relativity theory. For simplicity, we will only consider the case for which all velocities are along the same line of motion. We consider two inertial frames of reference A and B where B moves relative to A with a velocity v_{BA}. Now, let observers in A and B observe the motion of a point P. The velocity of point P as measured in A, v_{PA}, in terms of the velocity of P as measured in B, v_{PB}, is given by

$$v_{PA} = \frac{v_{PB} + v_{BA}}{1 + \dfrac{v_{PB}v_{BA}}{c^2}}$$

Keep in mind that the velocities in the above expression are one-dimensional vectors (not vector magnitudes), so that they can have negative values. Also recall that $v_{AB} = -v_{BA}$. The numerator in the above equation is the same as we used in Chapter 3; the effect of the denominator is that no velocity will become greater than the speed of light.

Exercise 29–3 Velocity Addition Rocket A moves directly away from you at a speed of $0.87c$. If rocket B is moving directly toward rocket A, head-on, along the same line of motion at $0.75c$ relative to A, how fast, and in what direction is rocket B moving relative to you?

Solution

Let the subscript Y represent you. Then, the velocity addition equation gives

$$v_{BY} = \frac{v_{BA} + v_{AY}}{1 + \frac{v_{BA} v_{AY}}{c^2}}$$

If we take v_{AY} to be in the positive direction, then v_{BA} is in the negative direction. Thus,

$$v_{BY} = \frac{-0.75c + 0.87c}{1 + \frac{(-0.75c)(0.87c)}{c^2}} = \frac{0.12c}{0.3475} = 0.35c$$

Therefore, rocket B is moving away from you at 35% of the speed of light.

Practice Quiz

5. An object moving away from you at 0.80 times the speed of light shines a light in the direction of its forward motion. How fast is the beam of light moving away from you?

   (a) $1.8c$ **(b)** $0.80c$ **(c)** c **(d)** $0.60c$ **(e)** $1.7c$

6. An object moving toward you at 0.80 times the speed of light shines a light in the direction away from you. How fast does the beam of light move away from you?

   (a) $1.2c$ **(b)** $0.20c$ **(c)** c **(d)** $0.60c$ **(e)** $0.80c$

29–5 Relativistic Momentum

In order for the law of conservation of momentum to be the same in all inertial frames of reference, we must generalize the expression for the momentum of an object to an expression that is consistent with the theory of relativity. We call this result the **relativistic momentum** of a particle

$$p = \frac{mv}{\sqrt{1 - v^2/c^2}}$$

Note that in this equation v is the speed of the particle and not the relative velocity between two frames of reference. You can see that for a speed that is small when compared with c, the denominator is nearly equal to 1, so that this relativistic momentum is completely consistent with the $p = mv$ that we normally use at low speeds. In fact, the expression $p = mv$ can be retained if we reinterpret the mass of an object to be

$$m = \frac{m_0}{\sqrt{1 - v^2/c^2}}$$

where m_0 is called the **rest mass**, and m is called the **relativistic mass**.

Practice Quiz

7. For a particle with a rest mass of 2.6 mg, what will be its relativistic mass if it moves at 70% of the speed of light?

   (a) 1.4 mg **(b)** 2.6 mg **(c)** 4.7 mg **(d)** 3.6 mg **(e)** 3.1 mg

29–6 Relativistic Energy and E = mc²

One of the most well-known results of the theory of relativity is that mass is a form of energy. The equivalence between mass and energy is represented by the famous equation

$$E = mc^2$$

where m is the relativistic mass of the object. Note that the above equation includes both the energy of motion (its kinetic energy) and the energy content associated only with its rest mass, called the **rest energy,**

$$E_0 = m_0c^2$$

The kinetic energy can be found by taking the difference between the total energy E and the rest energy E_0. This relativistic kinetic energy is given by

$$KE = \frac{m_0c^2}{\sqrt{1-v^2/c^2}} - m_0c^2$$

It can be shown that at low speeds this expression for the kinetic energy is completely consistent with the more familiar $KE = \frac{1}{2}mv^2$.

Example 29–4 Relativistic Energy and Momentum What is the linear momentum and kinetic energy of an electron that is moving at 65% of the speed of light?

Solution

The rest mass of an electron is 9.11×10^{-31} kg. The relativistic momentum is given by

$$p = \frac{m_0v}{\sqrt{1-v^2/c^2}} = \frac{(9.11\times10^{-31}\text{ kg})(0.65c)}{\sqrt{1-(0.65)^2}} = 7.79\times10^{-31}\text{kg}\cdot c$$

$$= 7.79\times10^{-31}\text{kg}\cdot(3.00\times10^{8}\text{ m/s}) = 2.3\times10^{-22}\text{ kg}\cdot\text{m/s}$$

The kinetic energy is

$$KE = \frac{m_0c^2}{\sqrt{1-v^2/c^2}} - m_0c^2 = m_0c^2\left(\frac{1}{\sqrt{1-v^2/c^2}}-1\right)$$

$$= \left(9.11\times10^{-31}\text{ kg}\right)\left(3.00\times10^{8}\text{ m/s}\right)^2\left[\frac{1}{\sqrt{1-(0.65)^2}}-1\right] = 2.6\times10^{-14}\text{ J}$$

In this Example, I called the mass its rest mass and used the symbol "m_0." You should be aware, however, that the concept of relativistic mass is not universally used, and many would just use "m" in the above equations. The key is whether the factor $1/\sqrt{1-v^2/c^2}$ is written explicitly. If it is, then m is, in fact, the rest mass m_0; if it is not, then m is the relativistic mass. Thus, you need to pay attention to how you, and others, write the relativistic expressions. Both ways of doing it are in common use.

Practice Quiz

8. What is the energy content of 1.0 kg of mass?
   (a) 3.0×10^8 J **(b)** 9.0×10^{16} J **(c)** 0 J **(d)** 1.0 J **(e)** 1.0×10^{16} J

29–7 – 29–8 The Relativistic Universe and General Relativity

The theory of relativity now affects everyday life for most of us. Relativity is needed for many important medical applications; it is used to understand processes that lead to alternative (to coal) forms of energy, and it has become important for the pinpoint accuracy of navigation on Earth using the Global Positioning System (GPS).

In the discussion of special relativity, we limited ourselves to frames of reference that have constant relative velocities. In formulating his **general theory of relativity**, Einstein argued that, with regard to the laws of physics, observers in accelerated frames of reference are identical to observers in gravitational fields. This apparent fact is called the **principle of equivalence**. Therefore, the general theory of relativity is equivalent to a new theory of gravity that replaces Newton's law of universal gravitation.

Among the many very interesting phenomena that studying the theory of general relativity has brought forth are the effect of gravity on light, and the ability to study the physics of **black holes**, and also the prediction of **gravity waves**. A black hole is an object of such high density, and therefore high gravitational field, that the escape velocity from it exceeds the speed of light. Therefore, no object can escape from a black hole. For a spherical object of mass M to be a black hole, its radius must be at most

$$R = \frac{2GM}{c^2}$$

Reference Tools and Resources

I. Key Terms and Phrases

inertial frame of reference a frame of reference in which Newton's laws of motion are obeyed

event a specific point in space at a specific time

time dilation the principle of relativity that moving clocks run slowly

proper time the time interval between events that occur at the same point is space

length contraction the principle that moving lengths are shortened along the direction of relative motion

proper length the length of an object as measured in the rest frame of the object

relativistic momentum the expression for the momentum of a particle that is consistent with the conservation of momentum and the first postulate of special relativity

rest energy the energy content of the mass of an object as measured in an inertial frame at rest with respect to the object

II. Important Equations

Name/Topic	Equation	Explanation
time dilation	$\Delta t = \dfrac{\Delta t_0}{\sqrt{1-v^2/c^2}}$	The dependence of time interval on the relative velocity between inertial frames of reference
length contraction	$L = L_0\sqrt{1-\dfrac{v^2}{c^2}}$	The dependence of length, along the direction of relative motion, on the relative velocity between inertial frames of reference
addition of velocities	$v_{PA} = \dfrac{v_{PB}+v_{BA}}{1+v_{PB}v_{BA}/c^2}$	The relativistic velocity addition equation for one-dimensional motion
relativistic momentum	$p = \dfrac{m_0 v}{\sqrt{1-v^2/c^2}}$	The expression for the momentum of a particle that is consistent with the theory of relativity
relativistic energy	$E = \dfrac{m_0 c^2}{\sqrt{1-v^2/c^2}}$	The expression of the equivalence between mass and energy.

Puzzle

ETHER,... OR

In an attempt to explain the null result of the Michelson-Morley experiment, a proposal was put forward that Earth is dragging ether, the medium that is carrying light waves, along with it. This explanation did not agree with the observation that light from distant stars comes in at an angle that depends on the velocity of Earth and that changes throughout the year. This phenomenon is called the *aberration of starlight.* See the figures: A (stationary telescope), B (moving telescope pointed at the star), and C (moving telescope adjusted for the aberration).

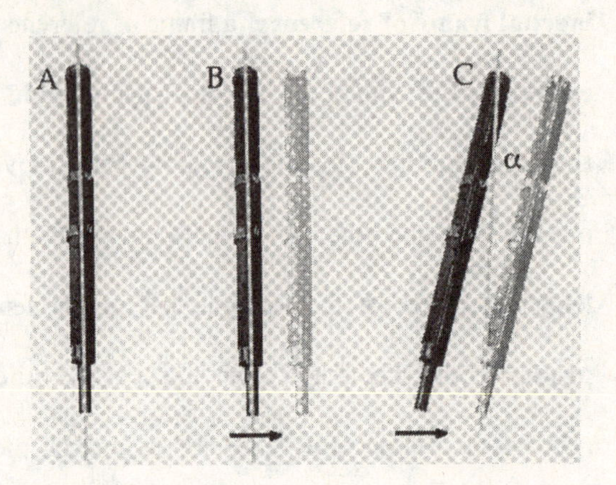

Give a rough estimate of the magnitude of the aberration angle, α. If the telescope tube was filled with a transparent substance of a high index of refraction, would you expect the aberration angle to change? If so, would it increase or decrease? Answer this question in words, not equations, briefly explaining how you obtained your answer.

Answers to Selected Conceptual Questions and Exercises

Conceptual Questions

2. The second postulate of relativity specifically refers to the speed of light *in a vacuum*. The speed of light in other mediums will always be less than the speed of light in a vacuum.

4. If the speed of light were only 35 mi/h, we would experience relativistic effects everyday. For example, a commuter would age more slowly than a person who works at home; a moving car would be noticeably shorter and distorted; you wouldn't be able to drive faster than 35 mi/h, no matter how powerful your car was and no matter how long you held the "pedal to the metal."

6. No, in both cases. The theory of relativity imposes no limits on the energy or momentum an object can have.

Conceptual Exercises

2. Recall that the proper time is the time between two events that occur at the same location, as seen by a given observer. **(a)** Proper time. **(b)** Proper time. **(c)** Proper time. **(d)** Dilated time.

8. The effects of length contraction – in fact, all relativistic effects – would be less than they are now if the speed of light were larger. In fact, in the limit of an infinite speed of light, there would be no relativistic effects at all.

10. In principle, the apple has a greater mass when near the top of its fall. The reason is that there is more gravitational potential energy in the Earth-apple system when the apple is at a greater height, and this increased energy is equivalent to an increased mass via the relation $E = mc^2$. See Conceptual Checkpoint 29-3 for a similar situation.

Solutions to Selected End-of-Chapter Problems

13. Picture the Problem: An astronaut measures her proper heartbeat at 72 beats per minute. Since the heart beat is related to the time, a person on Earth, which is traveling at a speed of 0.65c relative to the astronaut, will measure a time-dilated heartbeat.

Strategy: The heartbeat is a frequency (beats/minute) and frequency is the inverse of the period. Replace the time in Equation 29-2 with the inverse of the frequencies and solve the resulting equation for the frequency measured by an observer moving relative to the astronaut.

Solution: 1. (a) Since the time between each heartbeat will appear longer to the Earth-based observer, the measured heart rate will be $\boxed{\text{less than}}$ 72 beats per minute.

2. (b) Set the times in Equation 29-2 with the inverse of the frequency:

$$\Delta t = \frac{\Delta t_0}{\sqrt{1-v^2/c^2}}$$

$$1/f = \frac{1/f_0}{\sqrt{1-v^2/c^2}}$$

3. Solve for the heart beat frequency:

$$f = f_0\sqrt{1-v^2/c^2} = \left(72\,\tfrac{\text{beats}}{\text{min}}\right)\sqrt{1-(0.65)^2} = \boxed{55\,\tfrac{\text{beats}}{\text{min}}}$$

Insight: Since the frequency is the inverse of the period, as the time is dilated the frequency decreases.

25. Picture the Problem: An astronaut travels to a distant star and back at different speeds. The proper distance between the Earth and the star remains constant, but the contracted length on each leg of the trip differs since the astronaut travels at different speeds.

Strategy: Solve Equation 29-3 for the proper length. Set the Equation in terms of the outbound trip equal to the equation for the return trip and solve for the length of the return trip.

Solution: 1. Solve Equation 29-3 for the proper distance:

$$L = L_0\sqrt{1-v^2/c^2} \Rightarrow L_0 = \frac{L}{\sqrt{1-v^2/c^2}}$$

2. Set the proper distance on both trips equal and solve for the return distance:

$$L_0 = \frac{L_1}{\sqrt{1-v_1^2/c^2}} = \frac{L_2}{\sqrt{1-v_2^2/c^2}}$$

$$L_2 = L_1\frac{\sqrt{1-v_2^2/c^2}}{\sqrt{1-v_1^2/c^2}}$$

3. Insert the given data:

$$L_2 = 7.5\text{ ly}\frac{\sqrt{1-0.89^2}}{\sqrt{1-0.55^2}} = \boxed{4.1\text{ ly}}$$

Insight: Since the astronaut is traveling faster on the return trip, the distance is shorter. The proper distance for both trips is 9.0 ly.

39. Picture the Problem: The image shows two ships traveling toward the earth. Ship A has a speed $v_{AE} = 0.80c$ relative to the Earth. The speed of ship B relative to ship A is $v_{BA} = -0.50c$.

Strategy: We want to calculate v_{BE}, the speed of ship B relative to the Earth. Use Eq. 29-4 to get the speed.

Solution: 1. (a) Calculate the speed of ship B relative to the Earth:

$$v_{BE} = \frac{v_{BA}+v_{AE}}{1+v_{BA}v_{AE}/c^2} = \frac{(-0.50c)+0.80c}{1+(-0.50c)(0.80c)/c^2} = \boxed{0.50c}$$

2. (b) The relative speed increases by $\boxed{\text{more than}}$ $0.10c$ because increases in speed near the speed of light produce a smaller net increase than would be expected classically.

3. (c) Multiply both sides of Equation 29-4 by the denominator:

$$v_{BE} = \frac{v_{BA}+v_{AE}}{1+v_{BA}v_{AE}/c^2}$$

$$\therefore\ v_{BE}\left(1+v_{BA}v_{AE}/c^2\right) = v_{BA}+v_{AE}$$

4. Isolate the unknown variable v_{BA} and solve for the absolute value (the relative speed):

$$v_{BA}\left(1-v_{AE}v_{BE}/c^2\right) = v_{BE}-v_{AE}$$

$$\therefore\ |v_{BA}| = \left|\frac{v_{BE}-v_{AE}}{1-v_{BE}v_{AE}/c^2}\right| = \left|\frac{0.50c-0.90c}{1-(.50c)(0.90c)/c^2}\right| = \boxed{0.73c}$$

Insight: If ship A slowed by 0.10c, the relative speeds would have decreased by more than 0.10c to 0.31c.

45. Picture the Problem: The image shows two football players moving toward each other before a collision. After the collision the players stick together. In this problem the speed of light is 3.0 m/s.

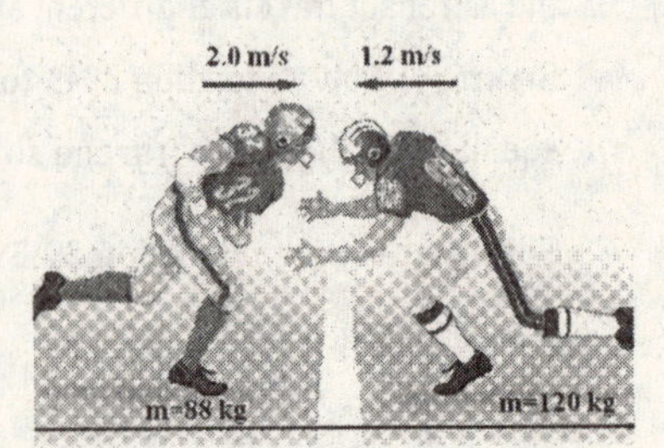

Strategy: We want to calculate the speed of the two players after the collision. Calculate the sum of the initial momentum using Equation 29-5. Set the initial momentum equal to the final momentum and solve for the final speed. Set the speed of light equal to 3.0 m/s for this problem.

Solution: 1. Set the initial momentum equal to the sum of the initial momentum: $p_i = p_1 + p_2 = \dfrac{m_1 v_1}{\sqrt{1-v_1^2/c^2}} + \dfrac{m_1 v_1}{\sqrt{1-v_1^2/c^2}}$

Solution: 1. Calculate the initial momentum: $p_i = \dfrac{88\text{ kg}(2.0\,\frac{\text{m}}{\text{s}})}{\sqrt{1-(2.0\,\frac{\text{m}}{\text{s}})^2/(3.0\,\frac{\text{m}}{\text{s}})^2}} + \dfrac{120\text{ kg}(-1.2\,\frac{\text{m}}{\text{s}})}{\sqrt{1-(1.2\,\frac{\text{m}}{\text{s}})^2/(3.0\,\frac{\text{m}}{\text{s}})^2}} = 79.01\text{ kg}\cdot\frac{\text{m}}{\text{s}}$

2. Set the final momentum equal to the initial and solve for the speed: $p_f = p_i = \dfrac{(m_1+m_2)v_f}{\sqrt{1-v_f^2/c^2}} \Rightarrow v_f = \dfrac{p_i}{\sqrt{(m_1+m_2)^2 + p_i^2/c^2}}$

$$\therefore\ v_f = \frac{79.01\text{ kg}\cdot\frac{\text{m}}{\text{s}}}{\sqrt{(88\text{ kg}+120\text{ kg})^2 + (79.01\text{ kg}\cdot\frac{\text{m}}{\text{s}})^2/(3.0\,\frac{\text{m}}{\text{s}})^2}} = \boxed{0.38\,\frac{\text{m}}{\text{s}}}$$

Insight: If this problem were worked classically, the final speed would be 0.15 m/s. The larger final speed is due to the increase in initial momentum of the first player due to relativistic effects.

61. Picture the Problem: The work required to increase the speed of a baseball from 25 m/s to 35 m/s is equal to the change in kinetic energy of the baseball at those speeds.

Strategy: Calculate the work using the work-energy theorem (Equation 7-7) with classical kinetic energy for the speeds of 25 m/s to 35 m/s. Use the work-energy theorem with relativistic kinetic energy (Equation 29-9) for the speeds 200,000,025 m/s to 200,000,035 m/s.

Solution: 1. (a) Calculate the work from Eq. 7-7:

$$W = \Delta K = K_f - K_i = \frac{1}{2}mv_f^2 - \frac{1}{2}mv_i^2 = \frac{1}{2}m\left(v_f^2 - v_i^2\right)$$
$$= \frac{1}{2}(0.145\text{ kg})\left[(35.0\text{ m/s})^2 - (25.0\text{ m/s})^2\right] = \boxed{43.5\text{ J}}$$

2. (b) Relativistic kinetic energies are greater than their classical counterparts, so the work required is $\boxed{\text{greater}}$.

3. (c) Calculate the work using the relativistic kinetic energies:

$$W = \Delta K = K_f - K_i = \left(\frac{m_0c^2}{\sqrt{1-v_f^2/c^2}} - m_0c^2\right) - \left(\frac{m_0c^2}{\sqrt{1-v_i^2/c^2}} - m_0c^2\right)$$
$$= m_0c^2\left(\frac{1}{\sqrt{1-v_f^2/c^2}} - \frac{1}{\sqrt{1-v_i^2/c^2}}\right)$$
$$= 0.145\text{ kg}\left(3.00\times10^8\,\tfrac{\text{m}}{\text{s}}\right)^2\left(\frac{1}{\sqrt{1-\left(\frac{2.00000035\times10^8\text{ m/s}}{3.00\times10^8\text{ m/s}}\right)^2}} - \frac{1}{\sqrt{1-\left(\frac{2.00000025\times10^8\text{ m/s}}{3.00\times10^8\text{ m/s}}\right)^2}}\right)$$
$$= \boxed{7.00\times10^8\text{ J}}$$

Insight: Using the classical kinetic energy for the work done in part (c) gives an incorrect work of $W = 2.90\times10^8$ J. This is significantly smaller than the actual work needed.

65. Picture the Problem: The radius of the black hole at the center of the galaxy is proportional to its mass.

Strategy: Calculate the Schwarzschild radius (Equation 29-10) using the mass a mass equal to 10^6 solar masses.

Solution: Set the mass in Equation 29-10 equal to 10^6 solar masses:

$$R = \frac{2GM}{c^2} = \frac{2\left(6.67\times10^{-11}\ \frac{\text{N}\cdot\text{m}^2}{\text{kg}^2}\right)10^6\left(2.00\times10^{30}\ \text{kg}\right)}{\left(3.00\times10^8\ \text{m/s}\right)^2} = \boxed{2.96\times10^6\ \text{km}}$$

Insight: This radius is on the order of the size of the sun.

77. Picture the Problem: The image shows a the rest length of a titanium rod as 22 meters while it makes an angle of 45° with respect to the horizontal. The image also shows the rod as the rocket travels with a speed of 0.97c. In this image the horizontal length is contracted.

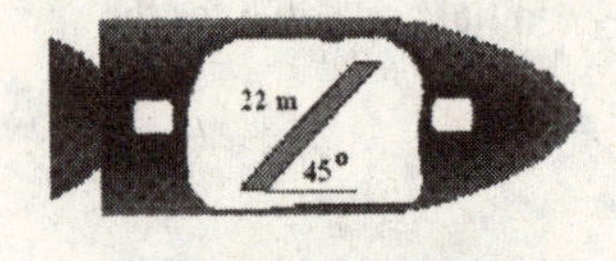

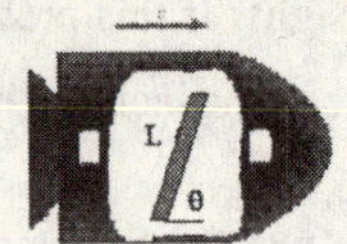

Strategy: Calculate the rest length of the horizontal and vertical components of the titanium rod using trigonometry. Then use Equation 29-3 to calculate the contracted length of the horizontal component. Finally, use the Pythagorean theorem and the definition of tangent on the contracted horizontal component and the vertical component to calculate the contracted length and angle of the moving titanium rod.

Solution: 1. Calculate the horizontal and vertical components of the rod's rest length:

$$L_{0x} = L_0 \cos\theta = (2.2\ \text{m})\cos(45°) = 1.556\ \text{m}$$

$$L_{0y} = L_y = L_0 \sin\theta = (2.2\ \text{m})\sin(45°) = 1.556\ \text{m}$$

2. Calculate the contracted horizontal component:

$$L_x = L_{0x}\sqrt{1-v^2/c^2} = (1.556\ \text{m})\sqrt{1-0.97^2} = 0.378\ \text{m}$$

3. (a) Calculate the contracted length:

$$L = \sqrt{L_x^2 + L_y^2} = \sqrt{(0.378\ \text{m})^2 + (1.556\ \text{m})^2} = \boxed{1.60\ \text{m}}$$

4. (b) Calculate the new angle:

$$\theta = \tan^{-1}\left(L_y/L_x\right) = \tan^{-1}(1.556\ \text{m}/0.378\ \text{m}) = \boxed{76°}$$

Insight: Since the horizontal length contracts, but the vertical length remains the same, the total length contracts and the angle of rotation increases.

Answers to Practice Quiz

1. (a) **2.** (b) **3.** (b) **4.** (a) **5.** (c) **6.** (c) **7.** (d) **8.** (b)

CHAPTER 30
QUANTUM PHYSICS

Chapter Objectives

After studying this chapter, you should

1. understand how the photon hypothesis explains blackbody radiation, the photoelectric effect, and Compton scattering.
2. be able to calculate the energy and momentum of a photon.
3. be able to determine the de Broglie wavelength of a particle.
4. know the results for the uncertainty principle for position and momentum and for energy and time.

Warm-Ups

1. In your own words, explain what happens in the photoelectric effect?
2. A certain radio signal has a wavelength as long as your arm. Estimate the energy of one photon of this radiation.
3. Estimate your de Broglie wavelength when running as fast as you can.
4. Estimate the uncertainty in the momentum of a hydrogen nucleus given that we know it is in a diatomic hydrogen molecule of length 0.075 nm.

Chapter Review

In the previous chapter, we discussed the theory of relativity as one of the principal corrections to Newton's laws that become most important for high speeds and strong gravitational fields. In this chapter, we discuss the other main correction, which becomes important at small size scales; we refer to these results as quantum physics.

30–1 Blackbody Radiation and Planck's Hypothesis of Quantized Energy

One of the first indications of the need for a theory of quantum physics came from the study of the electromagnetic radiation given off by a **blackbody**. An ideal blackbody is an object that absorbs all of the electromagnetic radiation that is incident upon its surface. This fact means that a blackbody is a perfect absorber of radiation, and in order to maintain thermal equilibrium, a blackbody is also a perfect radiator of electromagnetic energy. One of the most useful properties of blackbody radiation is that the distribution of energy given off only depends on the temperature of the object. In fact, the frequency at which a blackbody gives off the maximum intensity of electromagnetic radiation is directly proportional to the absolute temperature T

$$f_{peak} = \left(5.88\times10^{10}\,\text{s}^{-1}\cdot\text{K}^{-1}\right)T$$

This expression is known as **Wien's displacement law.**

The dependence of the intensity of blackbody radiation on frequency could not be accurately explained by classical physics. In order to reproduce the experimental result, Max Planck hypothesized that the radiant energy in a blackbody must be quantized in integral multiples of the constant h times the frequency:

$$E_n = nhf,\quad n = 1,2,3,\ldots$$

where the constant h is called **Planck's constant** and has the value

$$h = 6.63\times10^{-34}\,\text{J}\cdot\text{s}$$

Exercise 30–1 Wien's Displacement By how much is the frequency of the most intense light from a blackbody displaced if its temperature increases from 300 K to 380 K?

Solution

Using Wien's displacement law, we can see that

$$\begin{aligned} f_2 - f_1 &= \left(5.88\times10^{10}\,\text{s}^{-1}\cdot\text{K}^{-1}\right)\left[T_2 - T_1\right] \\ &= \left(5.88\times10^{10}\,\text{s}^{-1}\cdot\text{K}^{-1}\right)\left[80\text{ K}\right] = 4.7\times10^{12}\text{ Hz} \end{aligned}$$

At first sight this may look like an unreasonably large shift, but consider that in this temperature range the frequency of peak intensity is already on the order of 10^{13} Hz.

Practice Quiz

1. At what wavelength does a 450-K blackbody radiate energy with the most intensity?

   (a) 4.38×10^{13} m

   (b) 2.14×10^{-14} m

   (c) 6.41×10^{-6} m

   (d) 6.67×10^{5} m

   (e) None of the above.

30–2 Photons and the Photoelectric Effect

Planck's idea that electromagnetic radiation is quantized was considered to be a good explanation for blackbody radiation but not a general principle. It was Einstein who proposed that light, in general, comes in bundles of energy, called photons, that obey Planck's energy quantization result. Einstein argued that the energy of a photon is given by

$$E = hf$$

In addition to providing a more natural explanation of blackbody radiation, Einstein used his photon hypothesis to explain another phenomenon called the **photoelectric effect**. This effect occurs when light strikes the surface of a metal and ejects electrons. The minimum amount of energy needed to remove an electron from the surface of a metal is called the **work function**, W_0, for that metal. It was observed that no electrons are ejected from a metal unless the frequency of the incident light exceeds a certain **cutoff frequency**, f_0, given by

$$f_0 = \frac{W_0}{h}$$

It was further found that the maximum kinetic energy of the ejected electrons was independent of the intensity of the incident light and was only a function of its frequency:

$$KE_{\text{max}} = hf - W_0$$

Neither of these observations could be explained using classical physics. Einstein's photon hypothesis, however, provided a complete and accurate explanation of these results.

Example 30–2 Photons from the Sun The power output from the sun, called its luminosity is 3.826×10^{26} W. Estimate the number of photons per second that are leaving the surface of the Sun.

Picture the Problem The sketch shows the Sun with light emanating from its surface.

Strategy Since every photon carries a certain amount of energy, we need to determine how many photons will produce an amount of energy consistent with the Sun's luminosity.

Solution

1. The amount of energy coming from the Sun in 1 second is: $E = Pt = (3.826 \times 10^{26}\ \text{J/s})(1.00\ \text{s}) = 3.826 \times 10^{26}\ \text{J}$

2. Most of the Sun's energy is radiated near the middle of the electromagnetic spectrum, giving an effective frequency of:

$$f_{eff} = 5.45 \times 10^{14}\ \text{Hz}$$

3. The number, n, of photons at the above frequency that give the calculated energy is:

$$E = nhf_{eff} \quad \therefore$$

$$n = \frac{E}{hf_{eff}} = \frac{3.826 \times 10^{26}\ \text{J}}{\left(6.63 \times 10^{-34}\ \text{J} \cdot \text{s}\right)\left(5.45 \times 10^{14}\ \text{Hz}\right)}$$

$$= 1.06 \times 10^{45}$$

Insight As you might have expected, a large number of photons are given off in just 1 second.

Practice Quiz

2. If the wavelength of photon-1 is 20% longer than the wavelength of photon-2, then

   (a) the energy of photon-1 is 20% greater than the energy of photon-2.

   (b) the energy of photon-1 is 20% less than the energy of photon-2.

   (c) the energy of photon-2 is 20% greater than the energy of photon-1.

   (d) the energy of photon-2 is 20% less than the energy of photon-1.

   (e) both photons have the same energy.

3. What effect would increasing only the frequency of the light incident on a given metal have on the maximum kinetic energy of the electrons ejected from its surface?

   (a) It would have no effect on the maximum kinetic energy of the electrons.

   (b) It would decrease the maximum kinetic energy of the electrons by an amount directly proportional to the increase in frequency.

   (c) It would increase the work function of the metal, causing the maximum kinetic energy to decrease.

   (d) It would decrease the work function of the metal, causing the maximum kinetic energy to increase.

   (e) None of the above.

30–3 The Mass and Momentum of a Photon

Although photons are like particles in that they have energy and linear momentum, one way in which photons are not like ordinary particles is that they are massless. In fact, using the results of relativity in the previous chapter, it can be shown that any particle which travels at the speed of light must have zero rest mass.

However, as mentioned above, photons do carry linear momentum, and using the results of relativity, it can be shown that the momentum of a photon is given by

$$p=\frac{hf}{c}=\frac{h}{\lambda}$$

While the momentum of a typical photon is quite small, when large numbers of photons are incident on an object they can collectively provide a substantial *radiation pressure*.

Practice Quiz

4. If photon-1 has a longer wavelength than photon-2, then

(a) the momentum of photon-1 is greater than the momentum of photon-2.

(b) photon-1 has more energy than photon-2.

(c) the momentum of photon-1 is less than the momentum of photon-2.

(d) both photons have the same momentum.

(e) None of the above.

30–4 Photon Scattering and the Compton Effect

Another important verification of the photon idea was its ability to explain the **Compton effect**. The Compton effect is the result of how the direction and energy of light changes when scattered off electrons that are initially at rest. Using the photon idea, and the conservation of energy and momentum, an accurate explanation of the Compton effect emerged. Particularly, the shift in the wavelength of the scattered light is found to be

$$\Delta\lambda=\lambda'-\lambda=\frac{h}{m_e c}(1-\cos\theta)$$

where λ' is the wavelength of the scattered photon, λ is the wavelength of the incident photon, and θ is the angle at which the photon scatters relative to its incident direction (see Figure 30–7 in the text). The quantity $h/m_e c$ is called the Compton wavelength of an electron.

Practice Quiz

5. For the Compton scattering of a photon, does the greater shift in wavelength occur if the photon is scattered through 0° or 180°?

(a) 0° **(b)** 180° **(c)** They produce the same shift in wavelength.

(d) Neither angle produces any shift in wavelength.

(e) None of the above.

30–5 The de Broglie Hypothesis and Wave-Particle Duality

That light, which was traditionally understood to be a wave, has both wavelike and particle-like properties led Louis de Broglie to propose that matter, traditionally understood to be particles, may also exhibit wavelike behavior; we call this phenomenon **matter waves**. He postulated that the wavelength associated with a particle, such as an electron, would have the same relationship to its momentum as the wavelength of a photon has to its momentum

$$\lambda = \frac{h}{p}$$

The value of λ given by this expression is called the **de Broglie wavelength**. Notice that the particle-like nature (linear momentum) and the wavelike nature (wavelength) are intimately connected — you can't have one without the other (at a fundamental level).

Exercise 30–3 A Macroscopic Particle How fast must a particle of mass 1.00 g move to have a de Broglie wavelength of 1.00 mm?

Solution: We are given the following information.

Given: $m = 1.00$ g, $\lambda = 1.00$ mm; **Find**: v

The de Broglie wavelength is given by $\lambda = h/p$, where $p = mv$. Therefore,

$$\lambda = \frac{h}{mv} \quad \therefore \quad v = \frac{h}{m\lambda} = \frac{6.63\times10^{-34}\ \text{J}\cdot\text{s}}{\left(1.00\times10^{-3}\ \text{kg}\right)\left(1.00\times10^{-3}\ \text{m}\right)} = 6.63\times10^{-28}\ \text{m/s}$$

This velocity is essentially zero. At this speed, it would take an object an amount of time equal to the age of the universe to move a distance the size of an atom. What does this result say about seeing the effects of matter waves for macroscopic objects?

Verification of de Broglie's radical idea is found in the fact that beams of particles exhibit interference and diffraction patterns indicative of waves. Like X-rays, electrons and other particles scattered from crystalline materials produce interference patterns consistent with the constructive and destructive interference of waves scattering from different layers of the crystal. If d is the distance between adjacent layers of the crystal structure, and θ is the angle between the surface of the layer and the incident direction of the beam, then the condition for constructive interference is

$$2d\sin\theta = m\lambda, \quad m = 1,2,3,\ldots$$

Now we see that all matter and energy, in general, have both wavelike and particle-like properties. We call this phenomenon the **wave-particle duality**.

Practice Quiz

6. As the speed of a particle increases, its de Broglie wavelength

   (a) decreases. **(b)** increases. **(c)** stays the same. **(d)** None of the above.

7. What is the de Broglie wavelength of an electron moving at 90% of the speed of light?

   (a) 1.17×10^{-12} m **(b)** 5.64×10^{-22} m **(c)** 2.46×10^{-22} m **(d)** 2.29 m **(e)** 2.70×10^{-12} m

8. What is the de Broglie wavelength of a 0.50-kg ball thrown at 50.0 mi/h?

   (a) 2.7×10^{-35} m **(b)** 25 m **(c)** 1.3×10^{-35} m **(d)** 5.9×10^{-35} m **(e)** 11 m

30–6 The Heisenberg Uncertainty Principle

One of the most important lessons of quantum physics is that there is a fundamental uncertainty in nature that is related to the wave nature of matter; this fact is known as the **Heisenberg uncertainty principle**. One form of this principle states that both the position and momentum of a particle cannot be known with arbitrary precision at the same time. We can write this result in the following way:

$$\Delta p_y \Delta y \geq \frac{h}{2\pi}$$

where Δp_y is the uncertainty in the y component of momentum, and Δy is the uncertainty in the y component of position.

Another important expression of the uncertainty principle relates to energy and time:

$$\Delta E \Delta t \geq \frac{h}{2\pi}$$

In the above expression, ΔE is the uncertainty in the energy of a system, and Δt is the amount of time for some process to occur within the system.

Example 30–4 Uncertainty If the kinetic energy of an electron is measured to be 250 eV, and the uncertainty in its momentum is ±5.0%, what is the minimum uncertainty in its position?

Picture the Problem The sketch shows an electron moving in a certain direction that we'll call the y direction.

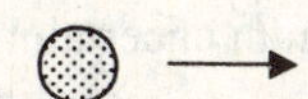

Strategy We need to determine its momentum so that we can find the uncertainty in it. From that information, we can use the uncertainty principle to get the minimum uncertainty in the position.

Solution

1. From the kinetic energy, we can determine the momentum:

$$KE = \frac{p^2}{2m_e} \quad \therefore \quad p = \sqrt{2m_e KE}$$

$$p = \sqrt{2\left(9.11\times10^{-31}\,\text{kg}\right)\left(250\,\text{eV}\right)\left(1.602\times10^{-19}\,\tfrac{\text{J}}{\text{eV}}\right)}$$

$$= 8.542\times10^{-24}\,\text{kg}\cdot\text{m/s}$$

2. The uncertainty in the momentum is from 5.0% less than the above result to 5.0% more:

$$\Delta p_y = 2(0.05)p$$

$$= 2(0.05)\left(8.542\times10^{-24}\,\text{kg}\cdot\text{m/s}\right)$$

$$= 8.542\times10^{-25}\,\text{kg}\cdot\text{m/s}$$

3. The minimum uncertainty in the position is given by the uncertainty principle:

$$\Delta p_y \Delta y_{\text{min}} = \frac{h}{2\pi} \quad \therefore$$

$$\Delta y_{\text{min}} = \frac{h}{2\pi \Delta p_y} = \frac{\left(6.63\times10^{-34}\,\text{J}\cdot\text{s}\right)}{2\pi\left(8.542\times10^{-25}\,\text{kg}\cdot\text{m/s}\right)}$$

$$= 1.24\times10^{-10}\,\text{m}$$

Insight Notice that when we used the uncertainty principle we used an equality instead of an inequality because we were seeking only the minimum uncertainty in position.

Practice Quiz

9. If the minimum uncertainty in an object's position is decreased by half, what can we say about the uncertainty in its momentum?
   (a) The uncertainty in momentum is at most half of what it was before the change.
   (b) The uncertainty in momentum is at least twice what it was before the change.
   (c) The uncertainty in momentum does not change.
   (d) The minimum uncertainty in momentum is precisely half of what it was before the change.
   (e) None of the above.

30–7 Quantum Tunneling

Still another fascinating phenomenon that was discovered as a result of quantum physics is that of quantum mechanical **tunneling**. This phenomenon is the fact that particles can be observed to emerge from across regions where they are forbidden to go according to classical physics. This effect is readily observed with light and has practical applications in devices such as the Scanning Tunneling Microscope (STM).

Reference Tools and Resources

I. Key Terms and Phrases

blackbody an object that absorbs all of the electromagnetic radiation incident upon it

Wien's displacement law the direct proportionality between the frequency at which a blackbody emits energy with maximum intensity and its absolute temperature

Planck's constant the fundamental constant of quantum physics

photoelectric effect the ejection of electrons by light incident upon the surface of a metal

work function the minimum amount of energy needed to remove an electron from the surface of a metal

Compton effect the scattering of a photon that collides with an electron at rest

matter wave the wavelike behavior of particles in accordance with the de Broglie relation

de Broglie wavelength the wavelength associated with a particle of matter

wave-particle duality the fact that matter and energy have both wavelike and particle-like properties

Heisenberg uncertainty principle the result that there is a fundamental uncertainty in nature

tunneling the result from quantum physics that particles can emerge from across regions where they are forbidden to go according to classical physics

II. Important Equations

Name/Topic	Equation	Explanation
Wien's displacement law	$f_{peak} = \left(1.04 \times 10^{11}\,\text{s}^{-1} \cdot \text{K}^{-1}\right)T$	The frequency at which a blackbody radiates with maximum intensity is directly proportional to the absolute temperature

photons	$E = hf$	The energy of a photon
	$p = \frac{hf}{c} = \frac{h}{\lambda}$	The momentum of a photon
de Broglie wavelength	$\lambda = \frac{h}{p}$	The wavelength of a particle of matter
uncertainty principle	$\Delta p_y \Delta y \geq \frac{h}{2\pi}$	The Heisenberg uncertainty principle for position and momentum
	$\Delta E \Delta t \geq \frac{h}{2\pi}$	The Heisenberg uncertainty principle for energy and time

III. Know Your Units

Quantity	Dimension	SI Unit
Planck's constant (h)	$[M][L^2][T^{-1}]$	J·s

Puzzle

UNCERTAINTY

An electron passes through your apparatus at $t = 0$. You measure its position with accuracy x_0. How accurately can you calculate its position later on, at time t_1? (Hint: The uncertainty in the electron's velocity is related to the uncertainty in its momentum.)

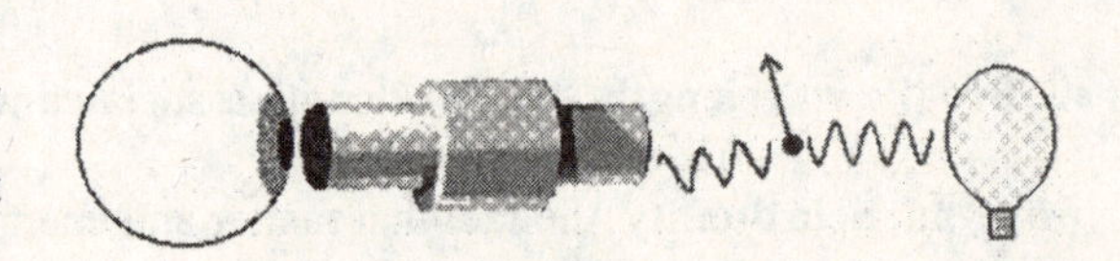

Answers to Selected Conceptual Questions and Exercises

Conceptual Questions

2. If energy is quantized, as suggested by Planck, the amount of energy for even a single high-frequency photon can be arbitrarily large. The finite energy in a blackbody simply can't produce such high-frequency photons, and therefore the infinite energy implied by the "ultraviolet catastrophe" cannot occur. In classical physics, any amount of energy can be in the form of high-frequency light – the energy does not have to be supplied in discrete, large lumps as in Planck's theory. Therefore, classical physics implies that all frequencies of light have the same amount of energy, no matter how high the frequency. This is what leads to the "catastrophe."

8. **(a)** A photon from a green light source *always* has less energy than a photon from a blue light source. **(b)** A photon from a green light source *always* has more energy than a photon from a red light source. The reason for these results is that the energy of a photon depends linearly on the frequency of light; that is, $E = hf$.

10. Classically, it should be possible to eject electrons with light of any frequency – all that is required is to increase the intensity of the beam of light sufficiently. The fact that this is not the case means that the classical picture is incorrect. In addition, the fact that there is a lowest frequency that will eject electrons implies that the energy of the photon is proportional to its frequency, in agreement with $E = hf$.

Conceptual Exercises

2. These sources have the same power; that is, they give off the same amount of energy per time. In addition, note that red photons have less energy than green photons, and green photons have less energy than blue photons. **(a)** The frequency of light increases as you go from red to green to blue, which means that the wavelength decreases. Therefore, the ranking in order of increasing wavelength is as follows: blue < green < red. **(b)** The ranking in order of increasing frequency is red < green < blue. **(c)** More red photons must be emitted to give the same energy as blue photons. Therefore, the ranking in order of increasing number of photons is blue < green < red.

6. Note that increasing the intensity while holding the frequency constant simply means that more photons – each with the original energy – strike the metal surface per second. **(a)** Stays the same. **(b)** Stays the same. **(c)** Increases. **(d)** Increases.

8. The scattered photon moves in the opposite direction to the incoming photon, which means that its scattering angle is 180°. To see this, suppose the initial photon moves in the positive x direction. This means that the initial y component of momentum is zero. After the collision, we are given that the electron moves in the positive x direction – it has no y component of momentum. Therefore, the scattered photon must also have zero y component of momentum, which means it will propagate in the negative x direction.

Solutions to Selected End-of-Chapter Problems

7. **Picture the Problem**: An incandescent light bulb operates at a cooler frequency than a halogen bulb. Therefore the blackbody radiation from the two light bulbs will have different peak frequencies.

Strategy: According to Wien's displacement law, the hotter the bulb the higher the peak frequency. Since the halogen bulb is hotter, it will have a higher frequency than the incandescent bulb. Use Wien's displacement law (Equation 30-1) to calculate both of the peak frequencies and calculate their ratio.

Solution: 1. (a) Since $f_{\text{peak}} \propto T$, the $\boxed{\text{halogen}}$ bulb has the higher peak frequency.

2. (b) Calculate the frequency of the halogen bulb: $f_{\text{hal}} = (5.88\times10^{10}\ \text{s}^{-1}\cdot\text{K}^{-1})(3400\ \text{K}) = 2.0\times10^{14}\ \text{Hz}$

3. Calculate the frequency of the standard bulb: $f_{\text{std}} = (5.88\times10^{10}\ \text{s}^{-1}\cdot\text{K}^{-1})(2900\ \text{K}) = 1.7\times10^{14}\ \text{Hz}$

4. Calculate the ratio of the two peak frequencies: $\dfrac{f_{\text{hal}}}{f_{\text{std}}} = \dfrac{1.999\times10^{14}\ \text{Hz}}{1.705\times10^{14}\ \text{Hz}} = \boxed{1.2}$

5 (c) The $\boxed{\text{halogen}}$ bulb produces a peak frequency closer to 5.5×10^{14} Hz than the standard incandescent bulb.

Insight: Since the halogen bulb produces a peak frequency closer to the sensitive frequencies of the human eye, more of the radiation from the halogen bulb is visible, making it appear brighter for the same power output.

21. Picture the Problem: A 150-Watt red light bulb emits photons with wavelength of 650 nm and a 25-Watt blue light blub emits photons with a wavelength of 460 nm.

Strategy: Use Equations 30-4 and 14-1 to calculate the energy of the photons for each bulb. Divide the power by the energy of each photon to calculate the rate at which photons are emitted.

Solution: 1. (a) The energy emitted per photon is inversely related to the wavelength, so the red light bulb must emit more photons to produce the same power output. The red bulb also produces more power than the blue bulb so the $\boxed{\text{red}}$ bulb emits more photons per second than the blue bulb.

2. (b) The energy per photon is inversely proportional to the wavelength. Since the blue bulb has the higher frequency, the $\boxed{\text{blue}}$ bulb emits photons of higher energy than the red bulb.

3. (c) Calculate the energy of the red photon: $E = hf = h\dfrac{c}{\lambda} = \left(6.63\times10^{-34}\ \text{J}\cdot\text{s}\right)\dfrac{3.00\times10^{8}\ \text{m/s}}{650\times10^{-9}\ \text{m}} = 3.06\times10^{-19}\ \text{J}$

4. Divide the power by the energy: $n_{\text{red}} = \dfrac{P}{E} = \dfrac{150\ \text{W}}{3.06\times10^{-19}\ \text{J}} = \boxed{4.9\times10^{20}\ \text{photons/s}}$

5. Calculate the energy of the blue photon: $E = \left(6.63\times10^{-34}\ \text{J}\cdot\text{s}\right)\dfrac{3.00\times10^{8}\ \text{m/s}}{460\times10^{-9}\ \text{m}} = 4.32\times10^{-19}\ \text{J}$

6. Divide the power by the energy: $n_{\text{blue}} = \dfrac{P}{E} = \dfrac{25\ \text{W}}{4.32\times10^{-19}\ \text{J}} = \boxed{5.8\times10^{19}\ \text{photons/s}}$

Insight: If the blue bulb were to produced the same number of photons per second as the red bulb, it would emit a power of 210 W.

37. Picture the Problem: The momentum of a photon is proportional to its frequency.

Strategy: Calculate the momentum of the two photons using Equation 30-11.

Solution: 1. (a) Because $f_{\text{blue}} > f_{\text{red}}$, and because $p \propto f$, $p_{\text{blue}} > p_{\text{red}}$. A photon of $\boxed{\text{blue}}$ light has the greater momentum.

2. **(b)** Calculate the momentum of the red photon: $p_{\text{red}} = \frac{hf}{c} = \frac{6.63\times10^{-34}\text{ J}\cdot\text{s}\left(4.0\times10^{14}\text{ Hz}\right)}{3.00\times10^{8}\text{ m/s}} = \boxed{8.8\times10^{-28}\text{ kg}\cdot\frac{\text{m}}{\text{s}}}$

3. Calculate the momentum of the blue photon: $p_{\text{blue}} = \frac{6.63\times10^{-34}\text{ J}\cdot\text{s}\left(7.9\times10^{14}\text{ Hz}\right)}{3.00\times10^{8}\text{ m/s}} = \boxed{1.7\times10^{-27}\text{ kg}\cdot\frac{\text{m}}{\text{s}}}$

Insight: Since the frequency of the blue photon is about double the frequency of the red photon, the momentum of the blue photon is about double the frequency of the red photon.

45. **Picture the Problem**: Two different photons are scattered back from collisions with free-electrons.

Strategy: Use Equation 30-15 to calculate the change in wavelength for a scattering of 180°. Calculate the percent difference by dividing the change in wavelength by the wavelength of each photon.

Solution: 1. (a) Since the change in wavelength does not depend on the wavelength of the photon $\boxed{\text{the change in wavelength is the same for both photons}}$.

2. **(b)** Since the change in wavelength is the same, the photon with the shorter wavelength, $\lambda = 0.030$ nm, will experience the greater percent change. So, the $\boxed{\text{X-ray photon}}$ experiences the greater percent change.

3. **(c)** Calculate the change in photon length: $\Delta\lambda = \frac{h}{m_e c}(1-\cos\theta)$

$$= \frac{6.63\times10^{-34}\text{ J}\cdot\text{s}}{9.11\times10^{-31}\text{ kg}\left(3.00\times10^{8}\text{ m/s}\right)}(1-\cos180°) = 0.00485\text{ nm}$$

4. Calculate the percent change in length of the visible photon: $\frac{\Delta\lambda}{\lambda} = \frac{0.00485\text{ nm}}{520\text{ nm}}100\% = \boxed{9.3\times10^{-4}\%}$

5. Calculate the percent change in length of the X-ray: $\frac{\Delta\lambda}{\lambda} = \frac{0.00485\text{ nm}}{0.030\text{ nm}}100\% = \boxed{16\%}$

Insight: Even though the change in wavelength is constant, the X-ray transfers more energy to the electron in order to conserve momentum in the collision. As the wavelength of the photon decreases, the transfer of energy becomes a larger portion of the total energy of the photon.

57. **Picture the Problem**: An electron and proton have the same de Broglie wavelength, which means that they must also have the same momentum. Since they have different masses, their kinetic energies will differ.

Strategy: Use Equation 30-16 to write the kinetic energy in terms of the de Broglie wavelength. Then divide the kinetic energy of the proton by the kinetic energy of the electron to calculate their ratio.

Solution: 1. (a) Since the proton and electron have the same wavelength, they will have the same momentum. The kinetic energy, $K = p^2/2m$, is inversely proportional to the mass and $m_e < m_p$. For identical momenta, an $\boxed{\text{electron}}$ has a greater kinetic energy than a proton.

2. **(b)** Write the kinetic energy in terms of the wavelength: $K = p^2/2m = h^2/2m\lambda^2$

3. Calculate the ratio of the kinetic energies:

$$\frac{K_e}{K_p} = \frac{h^2/2m_e\lambda^2}{h^2/2m_p\lambda^2} = \frac{m_p}{m_e} = \frac{1.673\times10^{-27}\text{ kg}}{9.11\times10^{-31}\text{ kg}} = \boxed{1836}$$

Insight: In this problem, we used the classical equation for the kinetic energy. As the speed approaches the speed of light, the ratio of kinetic energies decreases due to relativistic effects.

67. Picture the Problem: When an electron is known to be within a region with uncertainty of 0.15 nm, it will have a minimum uncertainty in momentum as required by the uncertainty principle. That minimum uncertainty requires the electron to have a finite kinetic energy.

Strategy: Use Equation 30-19 to calculate the minimum uncertainty in the momentum. Set this minimum uncertainty as the momentum of the electron in the kinetic energy equation ($K = p^2/2m$).

Solution: 1. (a) Solve Equation 30-19 for the minimum uncertainty in momentum:

$$\Delta p \Delta x \geq \frac{h}{2\pi}$$

$$\Delta p_{\text{min}} = \frac{h}{2\pi \Delta x} = \frac{6.63\times10^{-34}\text{ J}\cdot\text{s}}{2\pi\left(0.15\times10^{-9}\text{ m}\right)} = \boxed{7.0\times10^{-25}\text{ kg}\cdot\tfrac{\text{m}}{\text{s}}}$$

2. (b) Calculate the minimum kinetic energy:

$$K = \frac{p^2}{2m} = \frac{\left(7.0\times10^{-25}\text{ kg}\cdot\tfrac{\text{m}}{\text{s}}\right)^2}{2\left(9.11\times10^{-31}\text{kg}\right)\left(1.60\times10^{-19}\text{ J/eV}\right)} = \boxed{1.7\text{ eV}}$$

Insight: Decreasing the uncertainty in the position will increase the uncertainty in the momentum proportionately and increase the kinetic energy by the square of the decrease. For example, if the uncertainty were decreased by a factor of three (to 0.05 nm) the momentum would triple and the kinetic energy would increase by a factor of nine.

Answers to Practice Quiz

1. (c) **2.** (c) **3.** (e) **4.** (c) **5.** (b) **6.** (a) **7.** (a) **8.** (d) **9.** (b)

CHAPTER 31
ATOMIC PHYSICS

Chapter Objectives

After studying this chapter, you should

1. be familiar with some of the early models of the atom.
2. know how to determine which photons are in the emission spectrum of hydrogen.
3. understand Bohr's model of the hydrogen atom and its relation to de Broglie waves.
4. be able to specify the states of multielectron atoms in terms of the four quantum numbers.
5. be able to write out the electronic structure of multielectron atoms using the Pauli exclusion principle.
6. know the basic mechanisms for production of X-rays, laser light, and fluorescent and phosphorescent light.

Warm-Ups

1. What experiment suggested that atoms are not solid but that they may have a hard nucleus encircled by negative electrons?
2. A 5-mW laser pointer beam has a diameter of 3 mm. The wavelength of laser light is 680 nanometers. Estimate the number of photons per second that hit the laser spot on the wall.
3. Niels Bohr suggested that an atom emits light as an electron, orbiting the atom, moves closer to the nucleus. How did Bohr explain the fact that the spectra of chemical elements (such as hydrogen) consists of discrete colors rather than the whole continuous spectrum?
4. The human retina can respond to the light energy of a single photon! The wavelength of visible light is of the order of 500 nm. Compare the energy of a photon of visible light to that of a 100-g rifle bullet traveling at the speed of sound (340 m/s).

Chapter Review

In this chapter, we discuss the ideas that make up our modern understanding of the atoms. We focus primarily on the simplest atom, hydrogen. The ideas discussed in this chapter are the foundation of what is called *quantum physics*. The development of quantum physics (or *quantum mechanics*) revolutionized physics (and all physical sciences) in the 1900s.

31–1 Early Models of the Atom

The concept of the atom was originally proposed as the indivisible, fundamental quantity out of which things are made. The 1897 discovery of the electron, by J. J. Thomson, changed this belief by revealing that atoms must have internal structure. Thomson proposed what is called the "plum-pudding model" of the atom, in which negatively charged electrons are embedded in a nearly uniform distribution of positively charged matter.

A new picture of atomic structure emerged after Ernest Rutherford scattered positively charged *alpha particles* from a thin gold foil. The results of Rutherford's experiments suggested that the structure of atoms is more like a miniature solar system rather than like a plum pudding. Rutherford's model placed all the positive charge in an atom, and most of its mass, at a small, central location, called the **nucleus**, with the negatively charged electrons moving around the nucleus in orbits, similar to the planets orbiting the Sun. While the results of scattering experiments made Rutherford's model seem more reasonable, it was not consistent with experiments on the light given off by atoms, nor was it a stable structure according to Maxwell's electromagnetic theory.

31–2 The Spectrum of Atomic Hydrogen

Applying a large potential difference across a tube containing a low-pressure atomic gas causes the atoms of the gas to give off light. Passing this light through a diffraction grating separates the light into different wavelengths, producing a **line spectrum**. The process just described produces an *emission spectrum*. However, when light consisting of different wavelengths is passed through a gas, some of the wavelengths from this light will be absorbed by the gas, and the resulting light can then produce a line spectrum that is an *absorption spectrum*. Each atom has its own unique emission/absorption spectrum.

For hydrogen, the wavelengths that make up its emission spectrum are given by the formula

$$\frac{1}{\lambda} = R\left(\frac{1}{n'^2} - \frac{1}{n^2}\right) \qquad \begin{array}{l} n' = 1, 2, 3, \ldots \\ n = n'+1, n'+2, n'+3, \ldots \end{array}$$

The constant R, called the *Rydberg constant*, is $R = 1.097 \times 10^7\ \text{m}^{-1}$. The best-known wavelengths of this spectrum are those from the *Lyman series* ($n' = 1$), the *Balmer series* ($n' = 2$), and the *Paschen series* ($n' = 3$); however, there are an infinite number of series because there is no upper limit on n'.

Exercise 31–1 The Paschen Series Determine the range of wavelengths that make up the Paschen series of the hydrogen spectrum.

Solution

The Paschen series has $n' = 3$. Therefore, the values of n range from 4 to ∞. For $n = 4$ we have

$$\frac{1}{\lambda_4} = R\left(\frac{1}{3^2} - \frac{1}{4^2}\right) = \left(1.097\times10^7\ \text{m}^{-1}\right)\left(\frac{1}{3^2} - \frac{1}{4^2}\right) = 5.333\times10^5\ \text{m}^{-1}$$

$$\Rightarrow\quad \lambda_4 = 1.875\times10^{-6}\ \text{m}$$

For $n = \infty$ we have $1/n = 0$. Therefore,

$$\frac{1}{\lambda_\infty} = R\left(\frac{1}{3^2} - \frac{1}{\infty}\right) = \left(1.097\times10^7\ \text{m}^{-1}\right)\left(\frac{1}{3^2} - 0\right) = 1.219\times10^6\ \text{m}^{-1}$$

$$\Rightarrow\quad \lambda_\infty = 8.203\times10^{-7}\ \text{m}$$

Can you determine to which parts of the electromagnetic spectrum these wavelengths belong?

Practice Quiz

1. Which of the following is not a wavelength in the Balmer series?

   (a) 656.3 nm **(b)** 486.2 nm **(c)** 364.6 nm **(d)** 434.1 nm **(e)** 541.4 nm

31–3 – 31–4 Bohr's Model of the Hydrogen Atom and de Broglie Waves

To deal with the inconsistency between Rutherford's model and the experiments on the spectrum of hydrogen, Bohr refined Rutherford's model with four assumptions:

1. The electron in a hydrogen atom moves in a circular orbit around the nucleus.
2. The only circular orbits that are allowed are those for which the angular momentum of the electron is given by $L_n = n\dfrac{h}{2\pi}$, for $n = 1, 2, 3, \ldots$, where h is Planck's constant.
3. Electrons do not give off electromagnetic radiation while in an allowed orbit.
4. Electromagnetic radiation is given off, or absorbed, by an atom only when an electron changes from one allowed orbit to another. The frequency of the single photon that is emitted or absorbed is $\Delta E = hf$, where ΔE is the energy difference between the two allowed orbits.

Use of the centripetal force for the circular motion from Coulomb's law leads to an expression for the radii of the allowed orbits in a hydrogen atom, often called **Bohr orbits,**

$$r_n = \left(\frac{h^2}{4\pi^2 mke^2}\right)n^2, \quad n = 1,2,3,\ldots$$

where m is the mass of an electron, e is the elementary charge, and $k = 1/4\pi\varepsilon_0$ from Coulomb's law. The speed with which an electron moves in a Bohr orbit is given by

$$\upsilon_n = \frac{2\pi ke^2}{nh}, \quad n = 1,2,3,\ldots$$

The above results can be extended to any single-electron atom, such as singly ionized helium and doubly ionized lithium, by accounting for the charge of the additional protons in the nucleus. In general, we take a nucleus to contain Z protons, giving it a charge of $+Ze$; the quantity Z is called the **atomic number** of the nucleus. The above results for r_n and υ_n then become valid for single-electron atoms with heavier nuclei if we replace e^2 by Ze^2.

Each Bohr orbit has a certain amount of energy given by the sum of the kinetic and electric potential energies of the electron. For single-electron atoms, the total energies of the orbits are given by

$$E_n = -\left(\frac{2\pi^2 mk^2e^4}{h^2}\right)\frac{Z^2}{n^2} = -(13.6\text{ eV})\frac{Z^2}{n^2}, \qquad n = 1,2,3\ldots$$

The lowest possible energy of the atom corresponds to $n = 1$ and is called the **ground state** of the atom. Orbits with energies higher than the ground state ($n > 1$) are called **excited states**. Using the above energy states, Bohr's model predicts an emission spectrum for hydrogen ($Z = 1$) that is described by the equation

$$\frac{1}{\lambda} = \left(\frac{2\pi^2 mk^2e^4}{h^3c}\right)\left(\frac{1}{n_f^2} - \frac{1}{n_i^2}\right)$$

where c is the speed of light. This result is consistent with the empirical result quoted previously, and in addition, it tells us from what quantities the Rydberg constant R is derived.

In 1923, de Broglie provided some physical insight into Bohr's model by showing that Bohr's condition on the allowed angular momenta of the orbiting electrons was equivalent to a standing wave condition for the electrons' matter waves: $n\lambda = 2\pi r_n$. This success helped the idea of matter waves to be taken more seriously. The properties of matter waves are determined by what is called **Schrodinger's equation**; this equation is the fundamental equation of quantum mechanics. Today, the most common view of matter waves is that the amplitude of a particle's matter wave, at a given location and at a given time, is related to the probability that the particle is located at that point at that time.

Example 31–2 Ionized Helium What is the energy of the 1st excited state of singly ionized helium?

Picture the Problem The sketch is of the first two Bohr orbits around the nucleus of a singly ionized helium atom (not to scale).

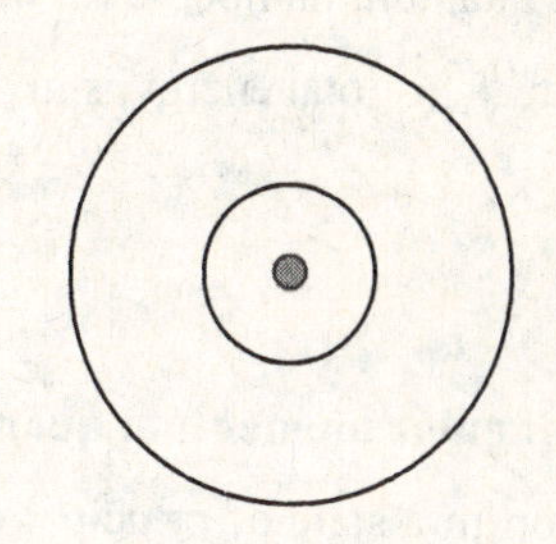

Strategy We use the more general expression for the allowed energies of single-electron atoms.

Solution

1. The atomic number of helium is: $Z = 2$

2. The 1st excited state is: $n = 2$

3. The energy then is: $E_2 = -(13.6\text{ eV})\dfrac{Z^2}{n^2} = -(13.6\text{ eV})\dfrac{2^2}{2^2} = -13.6\text{ eV}$

Insight This is the same energy as the ground state of hydrogen, but for a very different situation. The energy is negative because, as with planets orbiting the Sun, the electron and proton form a bound system, so that the (negative) potential energy "overpowers" the kinetic energy.

Practice Quiz

2. According to Bohr's model of hydrogen, what is the energy of an electron in the $n = 3$ excited state?

   (a) –6.04 eV **(b)** –1.51 eV **(c)** –13.6 eV **(d)** –0.85 eV **(e)** –4.53 eV

3. According to Bohr's model of hydrogen, what is the radius of the orbit of an electron in the $n = 3$ excited state?

   (a) 5.29×10^{-11} m **(b)** 1.59×10^{-10} m **(c)** 4.76×10^{-10} m **(d)** 1.06×10^{-10} m **(e)** 2.38×10^{-10} m

31–5 The Quantum Mechanical Hydrogen Atom

Aside from relativistic effects, the results from solving Schrodinger's equation for the hydrogen atom produce the best model of this atom that we have. From this analysis, we now describe the hydrogen atom using four parameters called *quantum numbers*.

The principal quantum number, n

The principal quantum number takes on all integer values, $n = 1, 2, 3, \ldots$, and determines the total energy of a given state. This total energy is given by, approximately, the same result produced by Bohr's model:

$$E_n = -(13.6\text{ eV})\frac{Z^2}{n^2}$$

The orbital angular momentum quantum number, ℓ

For an electron in a state of principal quantum number n, the orbital angular momentum of the electron can have only certain values as determined by the orbital angular momentum quantum number, ℓ, which takes on the values

$$\ell = 0,1,2,\ldots,(n-1)$$

The magnitude of the angular momentum of an electron with a given value of ℓ is

$$L = \sqrt{\ell(\ell+1)}\frac{h}{2\pi}$$

Notice that an orbiting electron is allowed to have an angular momentum of zero.

The magnetic quantum number, m_ℓ

When a hydrogen atom is in an external magnetic field, the allowed energy values of the electron, with specific values of n and ℓ, depend on an additional quantum number, m_ℓ. The range of values for this quantum number is from $-\ell$ to $+\ell$ in increments of 1:

$$m_\ell = -\ell, -\ell+1, -\ell+2, \ldots -1, 0, 1, \ldots, \ell-2, \ell-1, \ell$$

This magnetic quantum number specifies the component of the orbital angular momentum along a single direction; this direction is usually chosen to be the z-axis. The result is

$$L_z = m_\ell \frac{h}{2\pi}$$

One of the new consequences of the complete quantum theory is that only one component of the angular momentum can be known precisely at a given time.

The electron spin quantum number, m_s

This last quantum number results from the fact that electrons contain intrinsic angular momentum that is a property of the electron itself (like its mass and charge). This type of angular momentum is called the spin angular momentum of the electron. The quantum number m_s takes on two possible values:

$$m_s = -\frac{1}{2}, +\frac{1}{2}$$

When the values of all four quantum numbers are specified, we call that a particular **state** of the hydrogen atom. Notice that several different states can have the same energy. The solution of the Schrodinger equation suggests that when the atom is in a particular state, we should think of the matter wave of the electron in terms of a three-dimensional **probability cloud** instead of a specific path as suggested by Bohr's model. The probability cloud means that at any given time the electron has some finite probability of being anywhere around the nucleus.

Example 31–3 The Ground State of Hydrogen For an electron in the ground state of hydrogen, find **(a)** the total energy and **(b)** the magnitude of the orbital angular momentum. **(c)** Since it is possible to specify the value of only one component of the orbital angular momentum, what is that possible value for this atom?

Picture the Problem The sketch shows the Bohr orbit around the nucleus of the electron in a hydrogen atom.

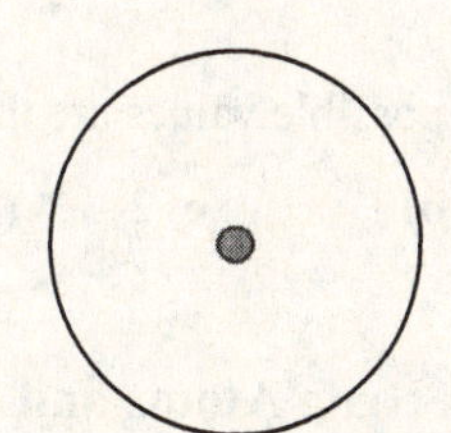

Strategy We must deduce the values of the relevant quantum numbers and use the above expressions to calculate the desired quantities.

Solution

Part (a)

1. The ground state of hydrogen has $Z = n = 1$, so the total energy is: $E_1 = -(13.6\text{ eV})\frac{1^2}{1^2} = -13.6\text{ eV}$

Part (b)

2. The range of the orbital angular momentum quantum number is: $\ell = 0 \ldots (n-1) \Rightarrow \ell = 0$

3. The magnitude of the orbital angular momentum then is: $L = \sqrt{\ell(\ell+1)}\frac{h}{2\pi} = 0 \cdot \frac{h}{2\pi} = 0$

Part (c)

4. The range of the magnetic quantum number is: $m_\ell = -\ell, \ldots, +\ell \Rightarrow m_\ell = 0$

5. The "z component" of the orbital angular momentum is: $L_z = m_\ell \frac{h}{2\pi} = 0 \cdot \frac{h}{2\pi} = 0$

Insight In part (b), $L = 0$ may seem odd for an orbiting electron, but keep in mind that the electron spends time in all directions at all distances. In part (c) the term "z component" is in quotes because it could really be any component. Because the atom is not in a magnetic field, no one component that we can call "z" is picked out.

Practice Quiz

4. How many possible values are there for ℓ when $n = 4$?

(a) 0 **(b)** 1 **(c)** 2 **(d)** 3 **(e)** 4

5. How many possible values are there for m_s when $n = 4$?

(a) 0 **(b)** 1 **(c)** 2 **(d)** 3 **(e)** 4

31–6 Multielectron Atoms and the Periodic Table

Unlike with hydrogen, the energies of the states of multielectron atoms depend on both n and ℓ. Electrons that have the same value of n are said to be in the same **shell.** The case of $n = 1$ is called the K shell, $n = 2$ is called the L shell, and so on, in alphabetical order. Electrons with the same value of ℓ are in the same subshell. The case $\ell = 0$ is called the s subshell, $\ell = 1$ is the p subshell, $\ell = 2$ is the d subshell, and $\ell = 3$ is the f subshell. After the f subshell, the names continue in alphabetical order.

Atoms naturally exist in their ground states unless some interactions are present to excite them. Finding the ground-state structure of multielectron atoms requires what is known as the **Pauli exclusion principle.** This principle states that

only one electron can occupy a given state of an atom.

This means that no two electrons can have the same four quantum numbers. Because of this principle, the ground state of an atom is obtained by filling up the allowed energy states, starting with the lowest (n = 1), until all the electrons are accounted for (see Example 31–4 below).

The **electronic configuration** of an atom is the specification of how many electrons exist in certain subshells. A particular notation for this specification lists the value of n, followed by the name of the subshell with the number of electrons in that subshell as a superscript. For example, the ground state of helium consists of two electrons, each with $n = 1$, $\ell = 0$, and $m_\ell = 0$, and one electron with $m_s = 1/2$, and the other with $m_s = -1/2$. The electronic configuration of this atom is $1s^2$. The electronic configuration of atoms leads to a better understanding of the **periodic table** of the elements. This table is a grouping of the chemical elements according to their properties. It is now known that elements with similar chemical properties correspond to elements with similar outer electron configurations.

Example 31–4 The Ground State of Boron The boron atom has an atomic number of 5; write out the electronic configuration of its ground state.

Solution

A neutral boron atom will have the same number of electrons as protons. Because its nucleus has an atomic number of 5, it also has 5 electrons surrounding the nucleus. For the $n = 1$ shell, ℓ can only be zero, corresponding to the s subshell. For the electrons in this 1s subshell, one can have $m_s = -1/2$ and the other can have $m_s = +1/2$, which places two electrons in this subshell, so the configuration starts with $1s^2$.

For the $n = 2$ shell, ℓ can only be either 0 or 1. There will be another two electrons in the s subshell ($\ell = 0$) for the same reason as before, so we write $2s^2$ for the configuration of these electrons. This leaves one electron unaccounted for. This last electron goes into the *p* subshell ($\ell = 1$). Therefore, the electronic configuration of the ground state of boron is

$$1s^2 2s^2 2p^1$$

Notice that for the $n = 2$ shell, the principal quantum number is indicated twice (for both the s and p subshells) which is redundant. Therefore, you should be aware that some authors would write this electronic configuration as $1s^2 2s^2 p^1$).

Practice Quiz

6. How many electrons can be placed in the L shell?

(a) 2 (b) 4 (c) 6 (d) 8 (e) 10

7. How many electrons can be placed in the d subshell?

(a) 2 (b) 4 (c) 6 (d) 8 (e) 10

8. Which of the following is the electronic structure of the ground state of nitrogen?

(a) $1s^2 2s^2 p^1$ (b) $1s^2 2s^2 p^2$ (c) $1s^2 2s^2 p^3$ (d) $1s^2 2s^2 p^4$ (e) $1s^2 2s^2 p^5$

31–7 Atomic Radiation

The spectra of multielectron atoms are more complicated than the spectrum of hydrogen. Three types of atomic radiation that are of practical use are discussed in this section. X-rays are high-energy photons that are often produced in tubes, called X-ray tubes, in which electrons are accelerated to high speeds and collided into a target. Most of the X-rays, called **bremsstrahlung**, are generated by the rapid deceleration of the electrons. Some of the X-rays are produced because of transitions of orbital electrons in the target atoms, which occur because the collision between the incident electrons and the target removes some of the lower-energy electrons in target atoms. When the states vacated by these electrons are filled, X-ray photons are given off.

The light given off by a **laser** is also due to interactions at the atomic level. With laser light, excited atoms are stimulated to emit radiation by incident light. The emitted photons then stimulate other excited atoms to emit photons, and so on. All these photons have the same energy and phase, and move in the same direction. This process is why they are called lasers, which is an acronym for **L**ight **A**mplification by the Stimulated **E**mission of **R**adiation.

Fluorescence occurs when atoms are illuminated by light and the energy of that light excites the atoms into a higher state. Once the illumination ceases, the electrons in the atoms spontaneously fall back to lower energy levels giving off (fluorescent) light of lower frequency than the light used to illuminate the atoms. A related phenomenon is **phosphorescence** in which the material continues to glow long after the original illumination ceases.

Reference Tools and Resources

I. Key Terms and Phrases

nucleus the central region of an atom that contains most of its mass and all of its positive charge

line spectrum the discrete spectrum of the wavelengths of light given off by atoms

Bohr orbit the circular orbits of electrons around the nucleus of an atom in Bohr's model

atomic number the number of protons in the nucleus of an atom

ground state the case when the electrons in an atom are in their lowest possible energy levels

excited state the case when one or more electrons in an atom are in energy levels above the ground state

principal quantum number an integer value that determines the total energy of a state of hydrogen and sets the boundaries for the values of other quantum numbers

orbital angular momentum quantum number an integer value that determines the orbital angular momentum of an electron in an atom and sets the boundaries for the magnetic quantum number

magnetic quantum number a quantity of integer value that determines the possible values of a component of the orbital angular momentum of an electron in an atom

electron spin quantum number a value that determines the intrinsic angular momentum of an electron

state of an atom the specification of the quantum numbers for every electron in an atom

probability cloud the interpretation of matter waves for the electrons around the nucleus of an atom

shell electrons with the same principal quantum number are in the same shell

subshell specification of the orbital angular momentum quantum number

Pauli exclusion principle the principle that only one electron can occupy a given state in an atom

electronic configuration specification of n, ℓ, and the number of electrons in each subshell

periodic table a table of the elements that organizes them by their chemical properties

laser light amplification by the stimulated emission of radiation

II. Important Equations

Name/Topic	Equation	Explanation
Bohr's model	$r_n = \left(\frac{h^2}{4\pi^2 mkZe^2}\right)n^2$ $n = 1,2,3,\ldots$	The radii of the Bohr orbits of single-electron atoms
	$E_n = -(13.6\text{ eV})\frac{Z^2}{n^2}$ $n = 1,2,3\ldots$	The allowed energy levels of single-electron atoms in Bohr's model
hydrogen spectrum	$\frac{1}{\lambda} = \left(\frac{2\pi^2 mk^2e^4}{h^3c}\right)\left(\frac{1}{n_f^2} - \frac{1}{n_i^2}\right)$	The wavelengths in the spectrum of hydrogen
orbital angular momentum	$L = \sqrt{\ell(\ell+1)}\frac{h}{2\pi}$	The magnitude of the orbital angular momentum from the associated quantum number

magnetic quantum number	$L_z = m_\ell \dfrac{h}{2\pi}$	The magnetic quantum number specifies one component of the orbital angular momentum

III. Know Your Units

Quantity	Dimension	SI Unit
Rydberg constant (R)	$[L^{-1}]$	m^{-1}

IV. Tips

The radius of the smallest Bohr orbit, for $n = 1$, is called the Bohr radius; it is typically denoted as a_0 and has a value of

$$a_0 = 5.29 \times 10^{-11}\ \text{m}$$

It is often convenient to write the expressions for the radii of the Bohr orbits, and their energies, in terms of a_0. These expressions become

$$r_n = n^2 \frac{a_0}{Z}$$

$$E_n = -\frac{ke^2}{2a_0}\frac{Z^2}{n^2}$$

where, again, $n = 1, 2, 3, \ldots$.

Puzzle

ISN'T THIS BACKWARD?

According to classical mechanics, as an electron shifts from an outer orbit to an inner orbit, its speed has to increase. Thus, the kinetic energy of the electron has to increase. Yet, according to Bohr, during such a process, a photon will take away the excess energy. Where does the excess energy come from?

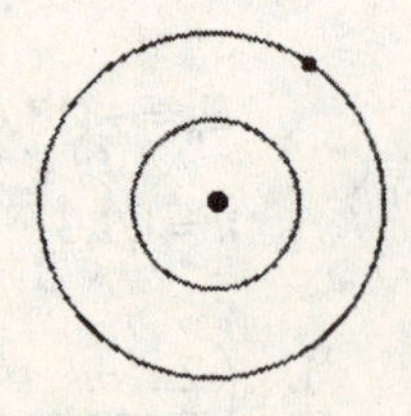

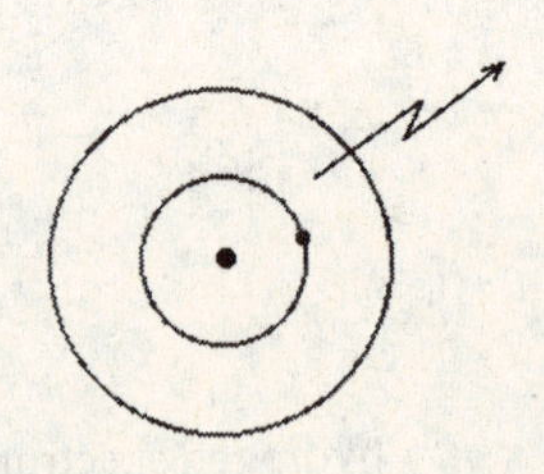

Answers to Selected Conceptual Questions and Exercises

Conceptual Questions

6. In principle, there are an infinite number of spectral lines in any given series. The lines become more closely spaced as one moves higher in the series, which makes them hard to distinguish in practice.

10. All of these questions can be answered by referring to Figure 31-17 and Table 31-3. **(a)** Not allowed; there is no *d* subshell in the $n = 2$ shell. **(b)** Not allowed for two reasons. First, there is no *p* subshell in the $n = 1$ shell. Second, a *p* subshell cannot hold 7 electrons. **(c)** Allowed. **(d)** Not allowed; the $n = 4$ shell does not have a *g* subshell.

12. No. Atoms in their ground states can emit no radiation. Even if an electron dropped from a highly excited state to the ground state in one of these atoms, the result would not be an X ray. The reason is that the binding energy of these atoms is simply much lower than the energy of a typical X-ray photon.

Conceptual Exercises

6. First, note that the radius of a Bohr orbit depends on n^2, and that the potential energy of the atom depends on $1/r$. The potential energy, then, depends on $1/n^2$. It follows that the potential energy in the $(n + 1)^{\text{th}}$ Bohr orbit is $Un^2/(n+1)^2$

8. **(a)** If an electron in the ground state absorbs a photon with an energy of 13.6 eV it will be on the verge of dissociation. Thus, 13.6 eV is the highest-energy photon the system can absorb. **(b)** The lowest-energy photon the system can absorb is one that would raise the electron from the $n = 1$ state ($E = -13.6$ eV) to the $n = 2$ state ($E = [-13.6 \text{ eV}]/4$). It follows that the lowest-energy photon is 3(13.6 eV)/4.

14. The wavelength of characteristic X-rays depends only on the type of atom in the target, and not on the energy of the incoming electrons. Therefore, these wavelengths stay the same.

Solutions to Selected End-of-Chapter Problems

7. **Picture the Problem**: The image shows the energy levels for the three longest wavelength transitions in the Lyman series. We want to find these wavelengths.

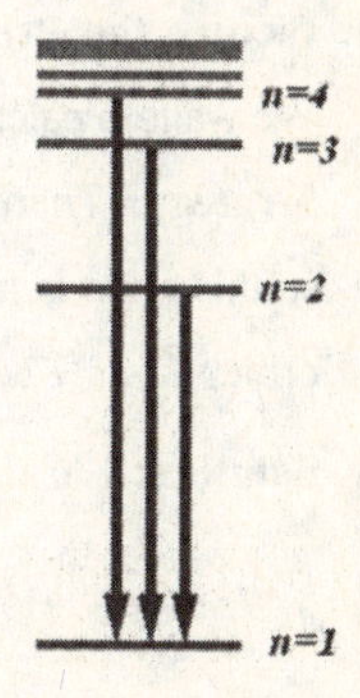

Strategy: Use Equation 31-2 to calculate the appropriate wavelengths. For the Lyman series set $n' = 1$. The longest wavelengths correspond to the smallest values of n. Therefore, set n equal to 2, 3, and 4 to obtain the longest wavelengths.

Solution: 1. Set $n' = 1$ in Equation 31-2 and solve for the wavelength:

$$\lambda = \frac{1}{R\left(1 - \frac{1}{n^2}\right)}$$

2. Set $n = 2$:

$$\lambda_2 = \frac{1}{\left(1.097\times10^{7}\ \text{m}^{-1}\right)\left(1-\frac{1}{2^2}\right)} = \boxed{121.5\ \text{nm}}$$

3. Set $n = 3$:

$$\lambda_3 = \frac{1}{\left(1.097\times10^{7}\ \text{m}^{-1}\right)\left(1-\frac{1}{3^2}\right)} = \boxed{102.6\ \text{nm}}$$

4. Set $n = 4$:

$$\lambda_4 = \frac{1}{\left(1.097\times10^{7}\ \text{m}^{-1}\right)\left(1-\frac{1}{4^2}\right)} = \boxed{97.23\ \text{nm}}$$

Insight: All three of these wavelengths lie in the ultraviolet portion of the electromagnetic spectrum.

19. Picture the Problem: The image shows the energy transition of an electron as it absorbs a photon and jumps from the n=3 to the n=5 orbit. It also shows the transition between the n=5 and n=7 orbits. We want to calculate the energy of the photon that will cause these transitions.

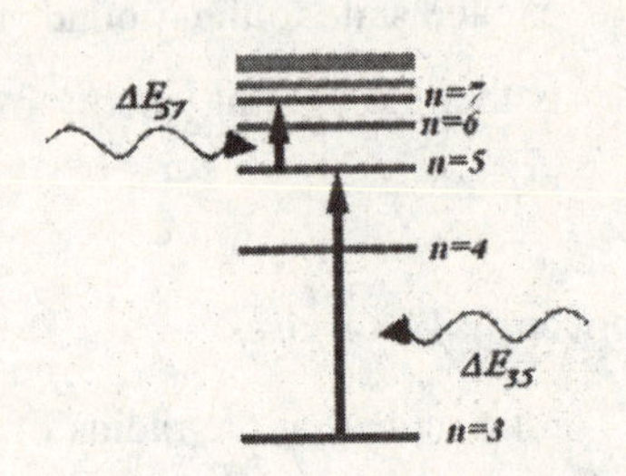

Strategy: The energy of the absorbed photon must be equal to the difference in the energies of the two electron levels. Calculate the difference in energy levels using Equation 31-9 (with Z=1).

Solution: 1. (a) Calculate change in energy: $|\Delta E| = |E_f - E_i|$

$$= \left|\frac{-13.6\ \text{eV}}{n_f^2} - \frac{-13.6\ \text{eV}}{n_i^2}\right| = 13.6\ \text{eV}\left(\frac{1}{n_i^2} - \frac{1}{n_f^2}\right)$$

2. Set $n_i = 3$ and $n_f = 5$:

$$|\Delta E_{35}| = 13.6\ \text{eV}\left(\frac{1}{3^2} - \frac{1}{5^2}\right) = \boxed{0.967\ \text{eV}}$$

3. (b) The energy of the photon would be $\boxed{\text{less than}}$ that found in part (a), since the absolute values of the lower n energy state differences are greater than those of the higher n energy state differences, and in both cases $\Delta n = 2$.

4. (c) Set $n_i = 5$ and $n_f = 7$:

$$|\Delta E_{57}| = 13.6\ \text{eV}\left(\frac{1}{5^2} - \frac{1}{7^2}\right) = \boxed{0.266\ \text{eV}}$$

Insight: The energy difference between levels rapidly decreases as n increases.

31. Picture the Problem: The image shows the Bohr orbit for a hydrogen atom. We want to calculate the radius of this orbit.

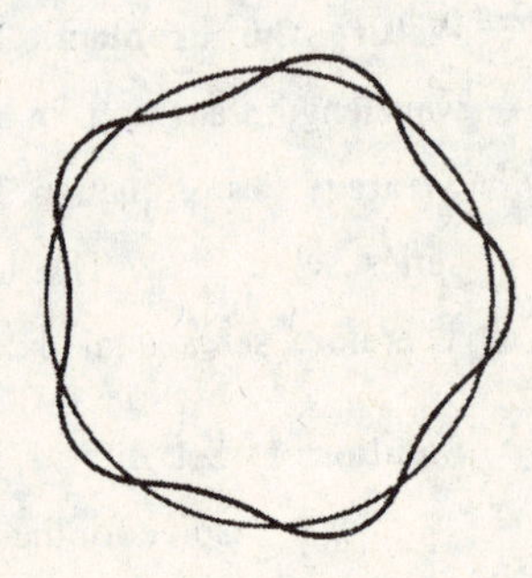

Strategy: There are five wavelengths contained in the orbit shown in the figure, so this atom is in the n=5 state. Use Equation 31-5 to calculate the radius of the orbit.

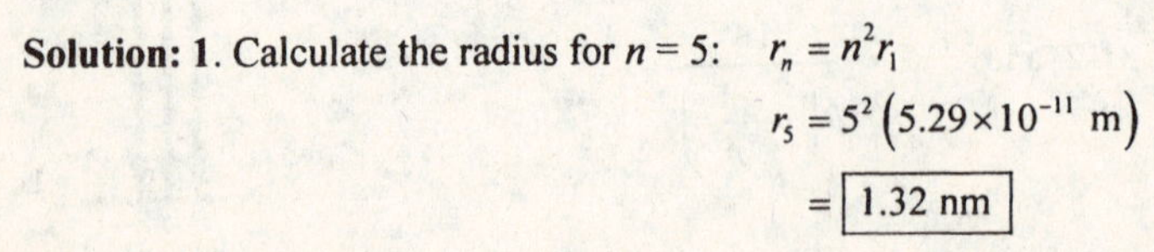

Solution: 1. Calculate the radius for $n = 5$: $r_n = n^2 r_1$

$$r_5 = 5^2\left(5.29\times10^{-11}\ \text{m}\right)$$

$$= \boxed{1.32\ \text{nm}}$$

Insight: The de Broglie wavelength is proportional to the orbit number and the number of wavelengths contained in the circumference is equal to the orbit number. These two facts require the radius to be proportional to the square of the orbit number.

37. **Picture the Problem**: We are given the magnitude of the orbital angular momentum and want to calculate the orbital quantum number and the minimum possible principal quantum number.

Strategy: Use Equation 31-12 to solve for the orbital quantum number, ℓ. Then, assume that this value of ℓ is the maximum for its given principal quantum number and use Equation 31-11 to calculate the minimum value of n. Use Equation 31-9 and the principal quantum number to determine the energy.

Solution: 1. (a) Insert the given value of L into Equation 31-12 and solve for the orbital quantum number:

$$\sqrt{\ell(\ell+1)}\,(h/2\pi)=10\sqrt{57}\,(h/2\pi)$$
$$\ell(\ell+1)=5700=75(76)\;\Rightarrow\;\ell=\boxed{75}$$

2. (b) Set the principal quantum number one larger than the orbital quantum number:

$$\ell_{max}=n-1$$
$$n=\ell_{max}+1=75+1=\boxed{76}$$

3. (c) Solve for the energy for n = 76:

$$E_{76}=-\frac{13.6\text{ eV}}{(76)^2}=\boxed{-2.35\text{ meV}}$$

Insight: To find the value of the integer ℓ, where $\ell(\ell+1)=X$ it is simplest to take the square root of X. The result lies between ℓ and $\ell+1$ since $\ell^2<\ell(\ell+1)<(\ell+1)^2$. Find the integer immediately below $\sqrt{X}$ and multiply it by the integer above to verify that the product is X.

47. **Picture the Problem**: Since the maximum orbital angular quantum number (ℓ) depends on the principal quantum number (n), the number of states possible in each shell will differ. We want to calculate the number of state possible for the principal quantum numbers n = 2, 3, and 4.

Strategy: For each ℓ, there are $(2\ell+1)$ values for m_ℓ and two values for m_s. Therefore for each ℓ there will be $2(2\ell+1)$ states possible. For each n the values of ℓ range from 0 to n–1. For n=2, sum the number of states possible for $\ell=0$ and $\ell=1$. For $n=3$, sum the states from $n=2$ with the added states for $\ell=2$. For $n=4$, sum the states from $n=3$ with the added states for $\ell=3$.

Solution: 1. (a) Sum the possible states for $\ell=0$ and $\ell=1$:

$$\text{total states }(n=2)=2[2(0)+1]+2[2(1)+1]$$
$$=2+6=\boxed{8}$$

2. (b) Add the possible states for $\ell=2$:

$$\text{total states }(n=3)=\text{total states }(n=2)+2[2(2)+1]$$
$$=8+10=\boxed{18}$$

3. (c) Add the possible states for $\ell=3$:

$$\text{total states }(n=4)=\text{total states }(n=3)+2[2(3)+1]$$
$$=18+14=\boxed{32}$$

Insight: For each higher value of n the number of states will increase because of the additional values of ℓ.

55. Picture the Problem: The image shows laser pulses directed toward the cornea during Photorefractive Keratectomy. The wavelength of the light is 193 nm. We want to calculate the difference in energy levels in the laser that produced this wavelength of light, and the number of photons required to deliver an energy of 1.58×10^{-13} J.

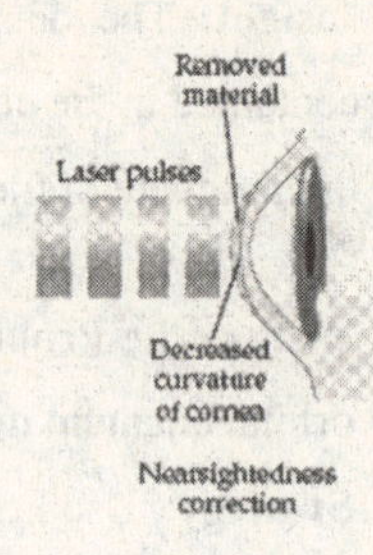

Strategy: Calculate the energy difference between the two levels by setting the energy difference equal to the energy of the photon using the relationship $|\Delta E| = hf = hc/\lambda$. Divide the total energy by the energy of each photon to calculate the number of photons needed.

Solution: 1. (a) Calculate the energy of the photon from the wavelength:

$$|\Delta E| = \frac{hc}{\lambda} = \frac{6.63\times10^{-34}\text{ J}\cdot\text{s}\left(3.00\times10^{8}\text{m/s}\right)}{193\times10^{-9}\text{m}}$$

$$= 1.03\times10^{-18}\text{J}\left(\frac{1\text{ eV}}{1.6\times10^{-19}\text{J}}\right) = \boxed{6.44\text{ eV}}$$

2. (b) Set the total energy equal to the energy per photon times the number of photons and solve for the number of photons:

$$E_{\text{total}} = nE_{photon}$$

$$n = \frac{E_{\text{total}}}{E_{\text{photon}}} = \frac{1.58\times10^{-13}\text{ J}}{1.029\times10^{-18}\text{ J}} = \boxed{1.53\times10^{5}\text{ photons}}$$

Insight: The photons from the laser interact with the molecules of the cornea heating them rapidly so they are vaporized off the surface, causing the cornea to flatten.

Answers to Practice Quiz

1. (e) **2.** (b) **3.** (c) **4.** (e) **5.** (c) **6.** (d) **7.** (e) **8.** (c)

CHAPTER 32

NUCLEAR PHYSICS and NUCLEAR RADIATION

Chapter Objectives

After studying this chapter, you should

1. know the basic constituents of the nucleus.
2. be familiar with the three processes of radioactive decay.
3. understand the exponential behavior of radioactive samples and know how radioactive dating works.
4. understand the effects of nuclear binding energy and why it leads to nuclear fission and fusion as energy sources.
5. know the difference between nuclear fission and nuclear fusion.
6. be aware of several practical applications of nuclear physics.

Warm-Ups

1. What aspect of short-lived radioactive substances makes them a health hazard? What aspect of long-lived radioactive substances makes them a health hazard?
2. The disintegration constant of technetium-99, an important medical tracer, is 10^{-13} per second. Estimate the half-life of technetium-99.
3. The activity of a radioactive sample depends on
   a. the property of the radioactive material and
   b. the size of the sample.

   The material is characterized by the disintegration (or decay) constant, which states what fraction of a sample will undergo the process per unit time. The disintegration constant of radioactive carbon-14 is equal to 3.84×10^{-12} per second. Based on that information, how many counts per minute would you expect for a mole of carbon-14? Explain your answer. (*Reminder*: A mole consists of 6×10^{23} items.)
4. The half-life of carbon-14 is 5730 years. Living organisms, such as plants, continuously replenish carbon-14 in their system from the atmosphere. After they die, the carbon-14 content is no longer maintained. You are given a piece of wood whose carbon-14 activity is 25% of what you observe in

living wood. How long has the wood been dead? (Explain in plain language how you obtained your answer by simple reasoning.)

Chapter Review

This final chapter primarily deals with nuclear and subnuclear physics. In this chapter, we touch on topics at the very forefront of human understanding. Nuclear physics is very important in everyday life from the standpoint of energy generation and even medical applications.

32–1 The Constituents and Structure of Nuclei

The nuclei of atoms consist of **protons** and **neutrons**; collectively these particles are called **nucleons**. The number of protons in a nucleus is called the **atomic number**, Z; the number of neutrons is called the **neutron number**, N. The total number of nucleons is called the **mass number**, A. Clearly,

$$A = Z + N$$

The chemical element to which a nucleus belongs is determined by the value of Z. Both Appendices E and F in your textbook list the atomic number of the chemical elements.

The notation used to specify the composition of the nucleus of a chemical element X is

$$^{A}_{Z}X$$

Sometimes Z is omitted because the value of Z is specified by the chemical element X. Even though Z is the same for every nucleus of a certain chemical element, the number of neutrons may not be the same. Nuclei having the same Z but different values of N are said to be different **isotopes** of the same nucleus. For example, $^{1}_{1}H$ and $^{3}_{1}H$ are different isotopes of the hydrogen nucleus with 0 neutrons (A = 1) and 2 neutrons (A = 3), respectively.

The masses of nuclei are often quoted in terms of the **atomic mass unit**, u. By definition,

$$1\ \text{u} = 1.660540 \times 10^{-27}\ \text{kg}$$

The mass of an atom, quoted in this unit, is often called its atomic mass. Appendix F in your textbook gives the atomic masses of many common isotopes. Because of the equivalence between mass and energy, $E = mc^2$, the mass of a nucleus is sometimes given in units of E/c^2: 1 u = 931.494 MeV/c^2. The size of a nucleus can be estimated by the empirical relationship

$$r = \left(1.2 \times 10^{-15}\ \text{m}\right) A^{1/3}$$

where A is the mass number. As this relationship shows, the radius of a nucleus is typically on the order of 10^{-15} m; this distance is called the **fermi** (fm).

A nucleus may contain many protons that are a very small distance apart. This situation leads to large electrostatic repulsion among the protons. Holding the nucleus together, against this repulsive force, is the **strong nuclear force.** This force is a fundamental force of nature that (a) is short range, acting only over a distance of a couple of fermi and (b) is attractive, acting with nearly equal strength among all nucleons. Because neutrons experience the strong nuclear force, but do not experience electrostatic repulsion (being electrically neutral), their presence in a nucleus helps to stabilize the nucleus (hold it together). The most stable nuclei are those with nearly equal numbers of protons and neutrons ($N \approx Z$). The more protons in a nucleus, the less stable it is; no nucleus with more than $Z = 83$ protons is stable.

Exercise 32–1 A Lithium Isotope Estimate the mass, in u, and the radius, in fm, of the isotope $^{7}_{3}\text{Li}$.

Solution

Appendix F in the textbook gives the atomic mass of a neutral lithium-7 atom to be 7.016005 u. To estimate the mass of the nucleus, we must subtract the mass of the $Z = 3$ electrons surrounding it. The mass of an electron is

$$m_e = 9.1094 \times 10^{-31}\,\text{kg}\left(\frac{1\,\text{u}}{1.660540 \times 10^{-27}\,\text{kg}}\right) = 5.4858 \times 10^{-4}\,\text{u}$$

Therefore,

$$M_{\text{Li-7}} = 7.016005\,\text{u} - 3\left(5.4858 \times 10^{-4}\,\text{u}\right) = 7.0144\,\text{u}$$

The given lithium isotope has $A = 7$. Therefore,

$$r_{\text{Li-7}} = \left(1.2 \times 10^{-15}\,\text{m}\right)A^{1/3} = \left(1.2 \times 10^{-15}\,\text{m}\right)7^{1/3} = 2.3 \times 10^{-15}\,\text{m} = 2.3\,\text{fm}$$

Practice Quiz

1. A particular isotope of carbon has 13 nucleons. How many neutrons are in the nucleus?

   (a) 5 (b) 6 (c) 7 (d) 8 (e) None of the above.

2. Approximately by what factor is the radius of $^{16}_{8}\text{O}$ greater than that of $^{8}_{4}\text{Be}$?

   (a) 2 (b) 8 (c) 1.26 (d) 1.59 (e) 1.41

32–2 Radioactivity

An unstable nucleus will disintegrate into a different nucleus; when it does it will emit one or more particles. Also, a nucleus in an excited state can make a transition to a lower energy state and emit a high-energy photon. Both of these processes are referred to as *nuclear decay,* and the emission that occurs is called **radioactivity**; therefore, nuclear decay is often called **radioactive decay**.

The main types of particles emitted during radioactive decay are:

* *alpha particles* (α), are helium nuclei ${}^{4}_{2}\text{He}$; these emitted particles are referred to as α-rays.
* *electrons*, called β^{-}-rays when emitted by nuclei.
* *positrons*, which are antielectrons having the same mass as an electron but with opposite charge, called β^{+}-rays when emitted by nuclei.
* *gamma rays* (γ), which are high-energy photons emitted when an excited nucleus decays to a lower energy state.

Radioactive decay that emits an alpha particle is called **alpha decay**. The initial unstable nucleus is called the **parent nucleus,** and the final nucleus is called the **daughter nucleus**. The daughter nucleus will have two fewer protons and two fewer neutrons than the parent nucleus. If the X represents the unstable parent, and Y is the daughter, we can write this process as

$${}^{A}_{Z}X \rightarrow {}^{A-4}_{Z-2}Y + {}^{4}_{2}\text{He}$$

Notice that the atomic and mass numbers on the left-hand side equals the sum of the corresponding atomic and mass numbers on the right-hand side.

Radioactive decay that emits a β-ray (either β^{+} or β^{-}) is called **beta decay**. In the beta decay emission of an electron, the basic process is that a neutron decays into a proton and an electron. Hence, the mass number remains the same, but the atomic number changes (increases) by 1,

$${}^{A}_{Z}X \rightarrow {}^{A}_{Z+1}Y + \text{e}^{-}$$

The process of β^{+} decay is more complicated to explain, but a similar result applies, except that the atomic number decreases by 1,

$${}^{A}_{Z}X \rightarrow {}^{A}_{Z-1}Y + \text{e}^{+}$$

When radioactive decay occurs, the total mass of the decay products (the final nucleus + emitted particles) is less than the mass of the initial nucleus. The difference in mass, Δm, results in a release of energy in the amount

$$E = |\Delta m| c^{2}$$

This fact can be used to predict how much kinetic energy the electron or positron should have as a result of beta decay. When the measured kinetic energies of the decay products in beta decay did not satisfy the conservation of energy, it was determined that another particle must be given off. This particle is called a **neutrino**. Neutrinos are very weakly interacting particles. A neutrino (ν) is given off in β^+ decay, and an antineutrino ($\overline{\nu}$) is given off during β^- decay.

Radioactive decay that emits a gamma-ray photon (γ) is called **gamma decay**. This process occurs when a nucleus in an excited state decays to a lower energy state. An excited nucleus is indicated by placing an asterisk as a superscript on the symbol. Thus, we have

$$^{A}_{Z}X^{*} \rightarrow {}^{A}_{Z}X + \gamma$$

Both A and Z remain the same during gamma decay.

The rate at which nuclear decay takes place is called the **activity**, R. A common unit of measure for activity is the **curie** (Ci), which is defined as

$$1 \text{ Ci} = 3.7 \times 10^{10} \text{ decays/s}$$

The SI unit of activity is the **becquerel** (Bq), which is defined as 1 Bq = 1 decay/s. Most commonly, activities are measured in millicuries (mCi) and microcuries (μCi).

Exercise 32–2 The Beta Decay of Carbon Estimate the energy released when carbon-14 undergoes β^- decay.

Picture the Problem In the diagram, the completely filled circles represent neutrons and the others are protons. The diagram shows the beta decay of a carbon-14 isotope.

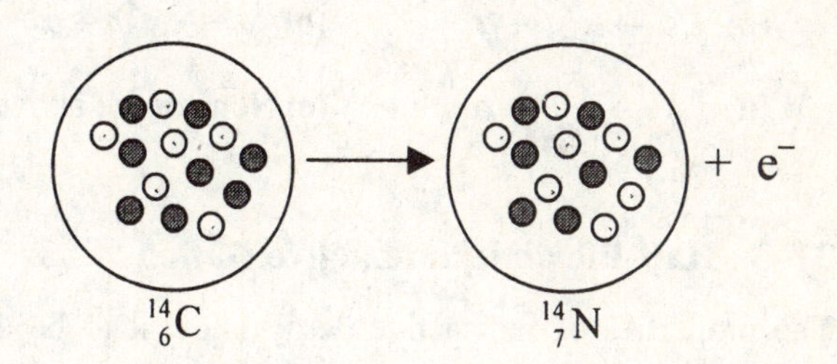

Strategy By comparing the masses of carbon-14 to that of nitrogen-14 plus an electron, we can estimate the energy released from the mass difference.

Solution

1. We can get the mass of a carbon-14 nucleus from Appendix F in the text: $M_C = 14.003242\,\text{u} - 6\left(5.4858\times10^{-4}\,\text{u}\right) = 13.999951\,\text{u}$

2. We can get the mass of a nitrogen-14 nucleus from Appendix F in the text: $M_N = 14.003074\,\text{u} - 7\left(5.4858\times10^{-4}\,\text{u}\right) = 13.999234\,\text{u}$

3. The mass difference is: $|\Delta M| = M_C - (M_N + m_e) = 13.99995\,\text{u} - (13.999783\,\text{u})$

$$= 0.000168\,\text{u}\left(\frac{931.494\,\text{MeV}/c^2}{\text{u}}\right) = 0.15649\,\text{MeV}/c^2$$

4. The energy released is: $E = |\Delta M|c^2 = \left(0.15649\frac{\text{MeV}}{c^2}\right)c^2 = 0.15649\text{ MeV}$

Insight Notice how many digits we had to keep in order to see the mass difference. Despite having to keep all these decimal places, the amount of energy is considerable. Another approach to doing this type of calculation is sometimes used. If you add the mass of 6 electrons to that of the carbon-14 nucleus, you have the mass of the carbon-14 atom. Also, add the mass of 6 electrons to that of the nitrogen-14 nucleus, and then, when you add the mass of the emitted electron, you get a total of 7 electrons, making the mass on the right-hand side that of the nitrogen-14 atom. Hence, the mass difference could be determined just from the differences in the atomic masses. This method allows you to do the problem in fewer calculations but the logic is less direct.

Practice Quiz

3. Which of the following represents a valid alpha decay process?

(a) ${}^{A}_{Z}X \rightarrow {}^{A}_{Z+1}Y + \alpha$ **(b)** ${}^{A}_{Z}X \rightarrow {}^{A}_{Z-1}Y + \alpha$ **(c)** ${}^{A}_{Z}X \rightarrow {}^{A-1}_{Z}X + \alpha$

(d) ${}^{A}_{Z}X \rightarrow {}^{A-4}_{Z-2}Y + \alpha$ **(e)** None of the above.

31–3 Half-life and Radioactive Dating

The properties of radioactive decay allow it to be used as a method of dating certain objects. This is because, for a given initial number of radioactive nuclei, N_0, the fraction of nuclei remaining at a given instant, N/N_0, depends, in a known way, on the amount of time t that has elapsed. This results in the well-known exponential decay equation

$$N = N_0 e^{-\lambda t}$$

where λ is called the **decay constant.** A large value of λ indicates a rapid decay process. A quantity that characterizes the speed of the decay process of a substance is its **half-life.** The half-life of a radioactive nucleus is the time interval required for the number of these nuclei to reduce by half. Using the above equation, we can determine that the half-life, $T_{½}$, is given by

$$T_{1/2} = \frac{\ln(2)}{\lambda}$$

The above behavior of radioactive nuclei results from the fact that the activity at a given time is directly proportional to the number of nuclei present at that time, with proportionality factor λ. Therefore, $R = \lambda N$. It then follows that

$$R = R_0 e^{-\lambda t}$$

If we measure the decay constant and present activity of a substance, then determine the initial activity of that substance by some means, we can approximately date the substance by solving the above equation for time:

$$t = -\frac{1}{\lambda}\ln\left(\frac{R}{R_0}\right)$$

In the case of carbon-14 dating of formerly living organisms, R_0 can be determined using the empirical fact that the ratio of carbon-14 to carbon-12 (which is not radioactive) tends to remain constant while organisms are alive. Therefore, we can measure the number of carbon-12 nuclei present today and assume it is the same as in the past. Knowing the ratio of carbon-14 to carbon-12 in living organisms, we can then find N_0 and set $R_0 = \lambda N_0$.

Example 32–3 Radioactive Decay A certain radioactive sample contains 5.00×10^{12} particles at a given instant. If its half-life is 12.2 minutes, how long will it take for 80% of the particles to decay?

Picture the Problem The sketch shows the sample at the initial instant, and after 80% has decayed.

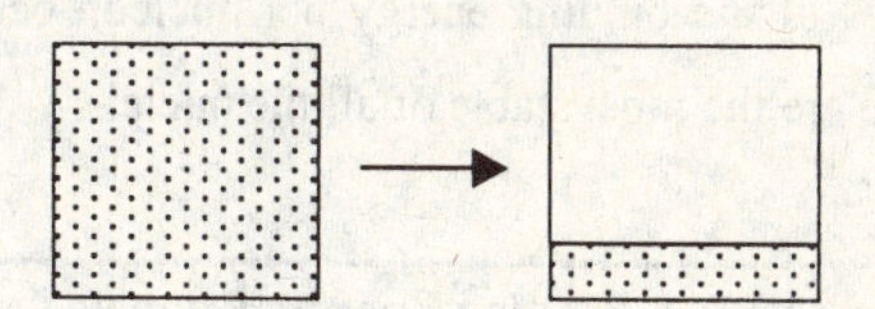

Strategy After 80% decays, only 20% is left. We can also use the half-life to determine the decay constant, giving us enough information to solve the problem.

Solution

1. Using that $N = 0.2N_0$, we can solve for t: $\quad 0.2N_0 = N_0 e^{-\lambda t} \;\Rightarrow\; t = -\frac{\ln(0.2)}{\lambda}$

2. The decay constant can be written as: $\quad \lambda = \frac{\ln(2)}{T_{1/2}} = \frac{\ln(2)}{12.2 \text{ min}} = 0.05682 \text{ min}^{-1}$

3. The time for 80% to decay becomes: $\quad t = -\frac{\ln(0.2)}{0.05682 \text{ min}^{-1}} = 28.3 \text{ min}$

Insight Verify the solution for t in step 1.

Practice Quiz

4. A certain radioactive nucleus, A, has a decay constant λ and a half-life T. If a different radioactive nucleus, B, has a decay constant of $\lambda/2$, the half-life of B is

   (a) $2T$ **(b)** T **(c)** $T/2$ **(d)** $4T$ **(e)** None of the above.

5. A certain radioactive nucleus, A, has a decay constant λ. A different radioactive nucleus, B, has a decay constant of $\lambda/2$. How do the activities compare at an instant when there are an equal number of nuclei of each type?

   (a) $R_B = 2R_A$ **(b)** $R_A = 2R_B$ **(c)** $R_A = R_B$ **(d)** $R_A = (2)^{1/3} R_B$ **(e)** None of the above.

31–4 Nuclear Binding Energy

An interesting fact is that the mass of all stable nuclei with more than one nucleon is less than the sum of the masses of the individual nucleons. This phenomenon occurs because of the **binding energy** of the nucleus. The binding energy is a measure of how tightly bound together the nucleons are; its magnitude equals the amount of energy required to break the nucleus apart into its constituent nucleons. This binding energy shows up as a reduction in mass, Δm, by the equivalence of mass and energy, $E = |\Delta m|c^2$. It is also useful to consider the binding energy per nucleon. An analysis of this quantity shows that the largest values of the binding energy per nucleon occur in the mass number range $50 < A < 75$. Nuclei in this range are the most stable of all the nuclei.

Exercise 32–4 Binding Energy Estimate the nuclear binding energy of ${}^{7}_{4}\text{Be}$.

Solution

The binding energy equals the energy needed to break apart the nucleus. Since $Z = 4$ and $A = 7$, breaking the nucleus produces 4 hydrogen atoms and 3 neutrons. So, the mass difference is determined by

$$\Delta M = ZM_H + Nm_n - M_{atom}$$

From Appendix F, the masses of hydrogen and beryllium are $M_H = 1.007825\text{ u}$ and $M_{Be} = 7.016930\text{ u}$. Therefore, we get

$$\Delta M = 4(1.007825\text{ u}) + 3(1.008665\text{ u}) - 7.016930\text{ u} = 0.040365\text{ u}\left(\frac{931.494\text{ MeV}/c^2}{\text{u}}\right) = 37.600\,\frac{\text{MeV}}{c^2}$$

Therefore, the binding energy 37.600 MeV.

Practice Quiz

6. Estimate the magnitude of the binding energy of deuterium in MeV/c^2.

   (a) 1.30 **(b)** 2.23 **(c)** 1.72 **(d)** 0.511 **(e)** None of the above.

32–5 Nuclear Fission

Under certain conditions, nuclei can break apart into smaller nuclei in a process called **nuclear fission.** The energy released in nuclear fission is very large compared to the energy released in chemical reactions. The fission of one nucleus can indirectly initiate the fission of other nuclei causing a **chain reaction**. The ability to control the chain reaction of $^{235}_{92}U$ has led to our modern use of nuclear power.

32–6 Nuclear Fusion

Given enough energy, light nuclei can combine to form a more massive nucleus in a process known as **nuclear fusion**. This larger nucleus has less mass than the individual nuclei that formed it. The mass difference is released as energy. As a potential energy source, nuclear fusion is even more powerful than fission. Because of the strong electric force of repulsion, or "Coulomb repulsion," among the nuclei when they are close together the nuclei must have a very high temperature. When fusion is initiated in a high-temperature gas containing nuclei, the process is called a **thermonuclear fusion reaction**. It requires a temperature of about 10^7 K to ignite thermonuclear fusion of protons (hydrogen nuclei) into helium nuclei. This latter process is called the proton-proton cycle and is the main energy source of the Sun.

32–7 Practical Application of Nuclear Physics

Biological Effects of Radiation

The high-energy particles given off by radioactive decay can be very damaging to living tissue by ionizing its molecules and therefore changing its structure. Several quantities have been defined in order to quantify the biological effects of radiation. The first of these quantities is the **roentgen** (R), which measures the amount of ionization caused by *X*-rays and γ-rays. A dosage of one roentgen of *X*-rays or γ-rays is the amount that produces 2.58×10^{-4} C of charge, from ionized molecules, in 1 kg of dry air at STP, that is,

$$1\text{ R} = 2.58 \times 10^{-4}\text{ C/kg}$$

Another quantity is the **R**adiation **A**bsorbed **D**ose, or **rad.** A dosage of 1 rad of any type of radiation delivered to a material is the amount of radiation that results in 0.01 J of absorbed energy by a 1-kg sample of material. So,

$$1\text{ rad} = 0.01\text{ J/kg}$$

Still another quantity is the **R**elative **B**iological **E**ffectiveness, or **RBE**. The *relative* in the RBE dose means relative to a 1-rad dose of 200-keV X-rays. A dose of 1 RBE of a particular type of radiation occurs when the ratio of the dose of 200 keV X-rays to the dose of the given type of radiation that would produce the same biological effect is 1. Thus, for a given biological effect,

$$\text{RBE} = \frac{\text{dose of 200 keV } X\text{-rays}}{\text{dose of given type of radiation}}$$

RBE is a dimensionless quantity. Another quantity that combines the rad and RBE is called the **biologically equivalent dose**. This dose is measured in a unit called **rem**, which stands for **R**oentgen **E**quivalent in **M**an. The definition of dosage in rem is

$$\text{dose in rem} = \text{dose in rad} \times \text{RBE}$$

With this definition, 1 rem of any type of radiation produces the same biological damage.

Magnetic Resonance Imaging

The interaction of nuclei with magnetic fields is used in magnetic resonance imaging, or MRI. An important advantage of MRI is that the radiation produced is low-energy radiation. Because the photons have low energy, they cause very little tissue damage.

PET Scans

In a Positron-Emission Tomography scan, or PET scan for short, a patient is given a radioactive substance that undergoes β^+ decay, giving off a positron. This positron almost immediately annihilates with an electron within the patient's body. The gamma-rays given off by this radiation penetrate through and leave the body where they can be detected during the scan. The analysis of the results from such a scan can provide important biological information.

32–8 Elementary Particles

One of the many goals of physics is to understand the world at its most fundamental level. This endeavor has led to the study of **elementary particles** that are thought to be the "building blocks" of all matter. To the best of our current understanding, elementary particles interact through four fundamental forces. These forces are called the strong nuclear force, the electromagnetic force, the **weak nuclear force**, and gravity. You have seen the need for each of these except the weak nuclear force; an explanation of this force is beyond the scope of this text, but the weak force is active in the process of beta decay.

Particles that experience the weak nuclear force but not the strong nuclear force are called **leptons**. There are six known leptons, of which the electron is one. As far as we know, no lepton has internal structure, so these are truly elementary particles. Particles that experience both the weak and strong nuclear forces are called **hadrons**; protons and neutrons fall into this category. All hadrons are made up of elementary particles called **quarks**. Some hadrons consist of two quarks, these are called **mesons**, and some hadrons consist of three quarks, these are called **baryons**. There are six quarks; these quarks are grouped into pairs called *flavors*. The flavors of the quarks are *up* (u) and *down* (d), *charm* (c) and *strange* (s), and *top* (t) and *bottom* (b). Quarks carry electric charges that are fractions of e, either $\pm(2/3)e$ or $\pm(1/3)e$, depending on the quark. Quarks also carry another kind of charge called *color*. There are three different quark colors usually referred to as red, green, and blue. The theory that describes the interaction of quarks via the color charge is called **quantum chromodynamics** (QCD).

32–9 Unified Forces and Cosmology

It is widely believed that at the beginning of the universe, at the Big Bang, there was only one fundamental force of nature, called the **unified force**. As the universe evolved, it is believed that the four forces we detect today separated from each other during processes that can be thought of as cosmological phase transitions. Today, scientists are trying to work backward to develop the theory of this unified force. One successful example is the **electroweak theory**, which has demonstrated that the electromagnetic and the weak nuclear forces are really just different aspects of the same basic force.

Reference Tools and Resources

I. Key Terms and Phrases

nucleon either a proton or a neutron in the nucleus

atomic number the number of protons in a nucleus

neutron number the number of neutrons in a nucleus

mass number the number of nucleons in a nucleus

isotopes the nuclei with the same number of protons but different numbers of neutrons

atomic mass unit a unit of mass, u, commonly used for atoms and nuclei: $1\ \text{u} = 1.660540 \times 10^{-27}$ kg

fermi a unit of distance, fm, commonly used in atomic and nuclear physics: $1\ \text{fm} = 10^{-15}$ m

strong nuclear force the fundamental force of nature that is responsible for holding nuclei together

radioactivity the high-energy emissions from nuclear decay due to a nucleus either being unstable or undergoing a transition from an excited state to a lower-energy state

activity the rate at which nuclear decay occurs

curie a common unit of measure for activity: 1 Ci = 3.7×10^{10} decays/s

half-life the time required for the number of radioactive nuclei to reduce by half

binding energy the energy that must be added to a nucleus to break it into its constituent nucleons

nuclear fission when a nucleus breaks apart into smaller nuclei

nuclear fusion when light nuclei combine to form a heavier nucleus

elementary particles the fundamental particles out of which matter is made

weak nuclear force a fundamental force of nature important in radioactive decay

leptons elementary particles that experience the weak force but not the strong force

hadrons particles that experience both the weak and strong nuclear forces and are made up of quarks

quarks fundamental particles that make up the hadrons

unified force the single force of nature believed to be at work at the beginning of the universe

II. Important Equations

Name/Topic	Equation	Explanation
nucleus	$r = \left(1.2 \times 10^{-15}\ \text{m}\right) A^{1/3}$	An empirical estimate of the radius of a nucleus
radioactivity	$N = N_0 e^{-\lambda t}$	The time dependence of the number of radioactive nuclei

III. Know Your Units

Quantity	Dimension	SI Unit
atomic number (Z)	dimensionless	—
neutron number (N)	dimensionless	—
mass number (A)	dimensionless	—
activity (R)	$[T^{-1}]$	Bq
half-life ($T_{1/2}$)	[T]	s
absorbed dose	$[L^2][T^{-2}]$	Gy
biologically equivalent dose	$[L^2][T^{-2}]$	Sv

IV. Tips

When it comes to the biological effects of radiation, the SI units are less familiar than those mentioned in the textbook. Nevertheless, you should be aware of what they are. For the absorbed dose of radiation, the SI unit is the gray (Gy): 1 Gy = 1 J/kg. For the biologically equivalent dose, the SI unit is the sievert (Sv): 1 Sv = 100 rem. These two units, the gray and the sievert, are referred to in the "Know Your Units" table.

Puzzle

A HOT HOUSE

Radium and radon are both naturally occurring radioactive substances found virtually everywhere. Why is radon considered a much more worrisome environmental problem?

Answers to Selected Conceptual Questions and Exercises

Conceptual Questions

2. The difference is that in an α decay only a single particle is emitted – the α particle – and it carries the energy released by the decay. In the case of β decay, two particles are emitted – the β particle (electron) and the corresponding antineutrino. These two particles can share the energy of decay in different amounts, which accounts for the range of observed energies for the β particles.

6. A change in isotope is simply a change in the number of neutrons in a nucleus. The electrons in the atom, however, respond only to the protons with their positive charge. Since electrons are responsible for chemical reactions, it follows that chemical properties are generally unaffected by a change in isotope.

8. Above the $N = Z$ line, a nucleus contains more neutrons than protons. This helps to make the nucleus stable, by spreading out the positive charge of the protons. If a nucleus were below the $N = Z$ line, it would have more protons than neutrons, and electrostatic repulsion would blow the nucleus apart.

Conceptual Exercises

4. The 14 decays in this series are as follows: α decay; β decay; β decay; α decay; α decay; α decay; α decay; β decay; β decay; α decay; α decay; β decay; α decay; β decay.

6. **(a)** Yes. **(b)** Yes. A given carbon-14 nucleus can decay at any time. All we can say for sure is that after 5730 years half of the initial carbon-14 nuclei (on average) will have decayed.

10. In two half lives, the activity of sample A will be reduced to one-quarter its initial value. The initial activity of sample B was half that of sample A, but after two half lives of sample A its activity is now one-quarter the initial activity of sample A – that is, the two samples have the same activity. It follows that the activity of sample B decreased by a factor of two in the same time that the activity of sample A decreased by a factor of four. Therefore, the half-life of sample B is twice as long as the half-life of sample A.

Solutions to Selected End-of-Chapter Problems

7. **Picture the Problem**: The image shows an alpha particle approaching a gold nucleus with a given initial kinetic energy. We want to calculate the speed of the alpha particle and its distance of closest approach.

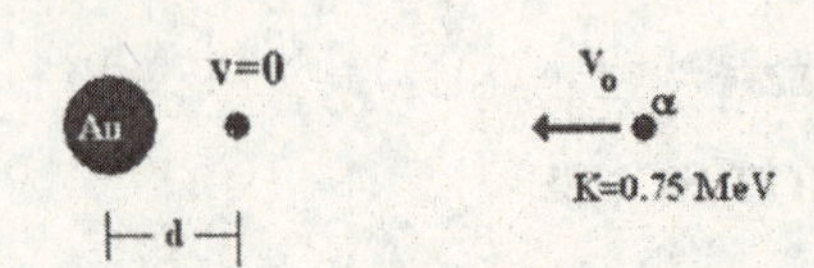

Strategy: Use Equation 7-6 to calculate the velocity from the kinetic energy. Set the initial kinetic energy equal to the electrostatic potential energy (Equation 20-8) at the closest approach and solve for the distance. Replace the nuclear charge of gold in the electrostatic potential energy equation with the charge on copper to calculate the distance of closest approach to the copper nucleus.

Solution: 1. (a) Solve Equation 7-6 for the velocity:

$$K = \frac{1}{2}mv^2 \Rightarrow v = \sqrt{\frac{2K}{m}}$$

2. Set the mass equal to the mass of the helium atom minus the mass of two electrons:

$$v = \sqrt{\frac{2(0.75\text{ MeV})}{4.002603\text{ u}\left(\frac{931.494\,\frac{\text{MeV}}{c^2}}{1\text{ u}}\right) - 2\left(0.511\,\frac{\text{MeV}}{c^2}\right)}}$$

$$= c\sqrt{\frac{1.5\text{ MeV}}{3727.38\text{ MeV}}} = \left(3.00\times10^8\,\tfrac{\text{m}}{\text{s}}\right)\sqrt{\frac{1.5}{3727.38}} = \boxed{6.0\times10^6\,\tfrac{\text{m}}{\text{s}}}$$

3. (b) Set the kinetic energy equal to the electrostatic energy and solve for the distance:

$$K = \frac{k(Ze)q}{d}$$

$$d = \frac{kZeq}{K} = \frac{\left(8.99\times10^9\text{ N}\cdot\text{m}^2/\text{C}^2\right)(79)(2)\left(1.60\times10^{-19}\text{ C}\right)^2}{0.75\times10^6\text{ eV}\left(1.60\times10^{-19}\text{ J/eV}\right)}$$

$$= \boxed{0.30\text{ pm}}$$

4. (c) Since $Z = 29$ for copper and $Z = 79$ for gold, the repulsive force the alpha particle experiences is much less when it approaches the copper nucleus. Thus, the distance of closest approach would be $\boxed{\text{less than}}$ that found in part (b).

Insight: The distance of closest approach for copper is 111 fm.

17. **Picture the Problem**: We are given two isotopes and told that they decay by alpha decay. We want to calculate the daughter isotopes and write an equation for each reaction. We also want to calculate the energy released in each reaction.

Strategy: For each parent isotope, determine the daughter isotope by subtracting the atomic number of the alpha particle (Z=2) from the atomic number of the parent. Also subtract the atomic mass of the alpha particle (A=4) from the atomic mass of the parent. From these results determine the daughter isotope and write the reaction equation. To calculate the energy released, subtract the masses of the daughter products (including the alpha particle) from the parent mass, where the masses are given in Appendix F. Multiply the mass defect by c^2 and convert the answers to MeV.

Solution: 1. (a) Determine the daughter isotope: $Z = 84 - 2 = 82$, which is Lead, or Pb.

$A = 212 - 4 = 208$

2. Write the reaction equation: $\boxed{{}^{212}_{84}\text{Po} \rightarrow {}^{208}_{82}\text{Pb} + {}^{4}_{2}\text{He}}$

3. Calculate the mass difference:

$$m_{\text{i}} = 211.988852 \text{ u}$$

$$m_{\text{f}} = 207.97664 \text{ u} + 4.002603 \text{ u} = 211.97924 \text{ u}$$

$$\Delta m = m_{\text{f}} - m_{\text{i}} = 211.97924 \text{ u} - 211.988852 \text{ u}$$

$$= -0.00961 \text{ u}$$

4. Multiply by c^2 and convert to MeV:

$$E = |\Delta m| c^2 = 0.00961 \text{ u}\left(\frac{931.494 \text{ MeV/c}^2}{1 \text{ u}}\right)c^2$$

$$= \boxed{8.95 \text{ MeV}}$$

5. (b) Determine the daughter isotope: $Z = 94 - 2 = 92$, which is Uranium, or U.

$A = 239 - 4 = 235$

6. Write the reaction equation: $\boxed{{}^{239}_{94}\text{Pu} \rightarrow {}^{235}_{92}\text{U} + {}^{4}_{2}\text{He}}$

7. Calculate the mass difference:

$$m_{\text{i}} = 239.052158 \text{ u}$$

$$m_{\text{f}} = 235.043925 \text{ u} + 4.002603 \text{ u} = 239.046528 \text{ u}$$

$$\Delta m = m_{\text{f}} - m_{\text{i}} = 239.046528 \text{ u} - 239.052158 \text{ u}$$

$$= -0.005630 \text{ u}$$

8. Multiply by c^2 and convert to MeV:

$$E = |\Delta m| c^2 = 0.005630 \text{ u}\left(\frac{931.494 \text{ MeV/c}^2}{1 \text{ u}}\right)c^2$$

$$= \boxed{5.244 \text{ MeV}}$$

Insight: The masses in Appendix F include the mass of the electrons in the atom. In this problem, the parent atoms had two more electrons than the daughter isotopes. These electrons were accounted for by using the mass of Helium (which contains two electrons) rather than the mass of the alpha particle (helium nucleus).

27. Picture the Problem: We are given the half-life of Tc-99 and want to calculate the decay constant and the number of nuclei necessary for a specific activity

Strategy: Use Equation 32-10 to calculate the decay constant from the half-life. Then solve Equation 32-11 for the necessary nuclei.

Solution: 1. (a) Calculate the decay constant: $\lambda = \frac{\ln 2}{T_{1/2}} = \frac{\ln 2}{6.05\text{ h}} = \boxed{0.115\text{ h}^{-1}}$

2. (b) Solve Equation 32-11 for the number of nuclei: $N = \frac{R}{\lambda} = \frac{1.50\times10^{-6}\text{ Ci}}{0.1146\text{ h}^{-1}}\left(\frac{3600\text{ s}}{\text{h}}\right)\left(\frac{3.7\times10^{10}\text{ s}^{-1}}{1\text{ Ci}}\right) = \boxed{1.7\times10^{9}\text{ nuclei}}$

Insight: This amount of Tc-99 would have a mass of only 2.8×10^{-13} g.

39. Picture the Problem: We want to calculate the amount of energy that must be added to an $^{16}_{8}\text{O}$ atom to eject one neutron, resulting in $^{15}_{8}\text{O}+^{1}_{0}\text{n}$.

Strategy: Calculate the difference in mass between the O-16 atom and the sum of the masses of an O-15 atom and a neutron. Calculating the energy by multiplying the mass defect by c^2 and converting the result to MeV.

Solution: 1. Calculate the mass of $^{15}_{8}\text{O}+^{1}_{0}\text{n}$: $m_f = 15.003065\text{ u} + 1.008665\text{ u} = 16.011730\text{ u}$

2. Subtract the mass of the $^{16}_{8}\text{O}$: $\Delta m = 16.011730\text{ u} - 15.994915\text{ u} = 0.016815\text{ u}$

3. Convert the mass to energy: $E = |\Delta m|c^2 = 0.016815\text{ u}\left(\frac{931.494\text{ MeV/}c^2}{1\text{ u}}\right)c^2 = \boxed{15.66\text{ MeV}}$

Insight: A neutron cannot spontaneously be ejected from $^{16}_{8}\text{O}$ without 15.663 MeV being added to the nucleus.

43. Picture the Problem: For the fission reaction, $^{1}_{0}\text{n}+^{235}_{92}\text{U}\rightarrow^{88}_{38}\text{Sr}+^{136}_{54}\text{Xe}+\text{neutrons}$, we want to calculate the number of neutrons produced and the energy released.

Strategy: Balance the atomic number and atomic mass on both sides of the reaction equation to identify the missing product. Calculate the difference in mass between the parent products $^{1}_{0}\text{n}+^{235}_{92}\text{U}$ and the daughter products $^{88}_{38}\text{Sr}+^{136}_{54}\text{Xe}+\text{neutrons}$. Convert this mass defect to energy to determine the amount of energy released in the reaction.

Solution: 1. Balance the atomic mass to determine the number of neutrons: $1+235 = 88+136+n(1)$

$n = 1+235-88-136 = 12$

2. Write out the complete reaction: $^{1}_{0}\text{n}+^{235}_{92}\text{U}\rightarrow^{88}_{38}\text{Sr}+^{136}_{54}\text{Xe}+\boxed{12\,^{1}_{0}\text{n}}$

3. Calculate the initial mass: $m_i = 1.008665\text{ u} + 235.043925\text{ u} = 236.052590\text{ u}$

4. Calculate the final mass: $m_f = 87.905625\text{ u} + 135.90722\text{ u} + 12(1.008665\text{ u}) = 235.91683\text{ u}$

5. Calculate the mass defect: $\Delta m = 235.91683\text{ u} - 236.052590\text{ u} = -0.13576\text{ u}$

6. Convert the mass defect to energy released: $E = |\Delta m|c^2 = 0.13576\ \text{u}\left(\dfrac{931.494\ \text{MeV/c}^2}{1\ \text{u}}\right)c^2$

$= \boxed{126.5\ \text{MeV}}$

Insight: This released energy is primarily carried away as kinetic energy of the neutrons.

51. Picture the Problem: We want to calculate the rate at which mass is converted to energy in the sun. Assuming the same rate of conversion over the lifetime of the sun, we want to calculate the total mass that has been converted to energy and what fraction of the sun's initial mass this represents.

Strategy: Divide the power output of the sun by the speed of light squared to calculate the mass per second converted to energy. Calculate the total mass converted to energy by multiplying the mass conversion rate by the age of the sun. Finally, to determine the percent of the sun's mass converted to energy divide the converted mass by the initial mass of the sun (the current mass plus the converted mass).

Solution: 1. (a) Calculate the rate of mass conversion:

$$\frac{\Delta m}{\Delta t} = \frac{P_{\text{Sun}}}{c^2} = \frac{3.90\times 10^{26}\ \text{W}}{(3.00\times 10^{8}\ \text{m/s})^2} = \boxed{4.33\times 10^{9}\ \text{kg/s}}$$

2. (b) Multiply the mass conversion rate by the age of the sun:

$$\Delta M = \frac{\Delta m}{\Delta t}T = (4.33\times 10^{9}\ \text{kg/s})(4.50\times 10^{9}\ \text{y})\left(\frac{3.16\times 10^{7}\text{s}}{1\ \text{y}}\right)$$

$$= 6.16\times 10^{26}\ \text{kg}$$

3. Divide the converted mass by the initial mass:

$$\frac{\Delta M}{M+\Delta M} = \frac{6.16\times 10^{26}\ \text{kg}}{2.00\times 10^{30}\ \text{kg} + 6.16\times 10^{26}\ \text{kg}}$$

$$= 3.07\times 10^{-4} = \boxed{0.0307\%}$$

Insight: The amount of mass converted by the sun to energy is negligible compared to the solar mass.

55. Picture the Problem: We want to calculate the energy absorbed by a 0.17 kg tumor when exposed to a radiation of 225 rad. We also want to calculate the resulting change in temperature of the tumor.

Strategy: Multiply the radiation dose by the mass of the tumor and convert the energy to joules. To calculate the change in temperature, solve Equation 16-13 for the temperature change, where $C = 4186\ \text{J/kg}\cdot\text{K}$.

Solution: 1. (a) The absorbed energy is:

$$E = (\text{Dose in rad})m = (225\ \text{rad})(0.17\ \text{kg})\left(\frac{0.01\ \text{J/kg}}{\text{rad}}\right) = \boxed{0.38\ \text{J}}$$

2. (b) Solve Eq. 16-13 for the temperature:

$$\Delta T = \frac{Q}{mc} = \frac{E}{mc} = \frac{0.3825\ \text{J}}{0.17\ \text{kg}(4186\ \text{J/kgK})} = \boxed{0.54\ \text{mK}}$$

Insight: The radiation does not significantly heat up the tumor.

Answers to Practice Quiz

1. (c) **2.** (c) **3.** (d) **4.** (a) **5.** (b) **6.** (b)